# Uncommon Cause

# Uncommon Cause

*Living for Environmental Justice in Kerala*

John Mathias

UNIVERSITY OF CALIFORNIA PRESS

University of California Press
Oakland, California



Parts of chapter 2 were adapted from John Mathias, "Scales of Value: Insiders and Outsiders in Environmental Organizing in South India," *Social Service Review* 91, no. 4 (2017), 621–651.

Parts of chapter 4 were adapted from Charles H. P. Zuckerman and John Mathias, "The Limits of Bodies: Gatherings and the Problem of Collective Presence," *American Anthropologist* 124, no. 2 (2022), 345–357.

Cataloging-in-Publication data is on file at the Library of Congress.

ISBN 978-0-520-39550-3 (cloth)
ISBN 978-0-520-39551-0 (pbk.)
ISBN 978-0-520-39552-7 (ebook)

33 32 31 30 29 28 27 26 25 24
10 9 8 7 6 5 4 3 2 1

*For Jali, Yoshi, and all those who dance in the rain*

# Contents

List of Illustrations *ix*
Acknowledgments *xi*
Note on Translation and Orthography *xv*

Introduction: Activist Lives *1*
1. Living for the People *29*
2. Living for Our People *62*
3. Uncommon Subjects *99*
4. Unquiet Objects *126*
Conclusion: Life Beyond Activism *161*

Appendix: Note on Methods *181*
Notes *187*
Bibliography *227*
Index *249*

# Illustrations

FIGURES


1. Faiza's nature camp / *4*
2. Evening gathering in Manamur protest tent / *8*
3. Rain camp / *30*
4. Dialogue Journey / *33*
5. Discussion in the *Kēraḷīyam* office / *40*
6. Editing in the *Kēraḷīyam* office / *40*
7. Protestors at the factory gates / *67*
8. Protestors inspecting truck / *70*
9. Protestors and politician in protest tent / *74*
10. Ingold's "Two views of the environment" / *89*
11. Solidarity view of local activism / *92*
12. Local view of solidarity activism / *94*
13. DuBois's "Stance triangle" / *108*
14. One World University / *109*
15. Protestor in polluted rice paddy / *131*
16. District Collector touring factory's pollution / *133*
17. Protestors checking news about themselves / *145*
18. Communist Party of India (Marxist) protest tent / *146*
19. Torchlit march / *149*

20. Arrest of fasting protestor / *154*
21. Faiza's choice in two triangles / *166*
22. Touring a large quarry / *171*
23. Loading a truck at Ahmed's uncle's quarry / *173*

MAPS

1. Map of Kerala / *16*
2. Map of Kerala's people's protests / *50*

# Acknowledgments

A core argument of this book is that a purpose is a kind of relationship with others. As I take the last steps toward the manuscript's completion (an aim pursued for many years now), I find that I have been living this argument. I look back on a long trail of conversations and exchanges—especially, of requests made and debts incurred. Here, I note a few relationships without which this manuscript would not be what it has become.

Funding for this research came from the Fulbright-Hays Doctoral Dissertation Research Abroad Fellowship, several institutions at the University of Michigan (the Rackham Graduate School, the Anthropology Department, the School of Social Work, the International Institute, and the Center for International and Comparative Studies) and the Center for Research and Creativity at Florida State University. The Center for Development Studies (CDS) in Ulloor provided research affiliation, use of their library, and occasional accommodation in their lovely guesthouse. The Center for Research, Education, and Social Transformation (CREST) provided institutional support for an early internship with the Malabar Coastal Institute for Training, Research, and Action (MCITRA), as well as ongoing guidance and feedback regarding this research. The American Institute of Indian Studies (AIIS), the US Department of Education, and the Big Ten Academic Alliance (BTAA) financially and institutionally supported my training in Malayalam. I will always be grateful to Nisha Kommatom for introducing me to Malayalam, and V.K. Bindu, Syam, Prema, and Arun for the boundless optimism with which they pursued my further instruction. Those lessons opened the door to everything described here.

I hope that the stories and ideas presented here will honor the friendships that made them possible. I am grateful to Sharath, Zabna, Yoshi, Bashirmaash, Subna, Rajinesh, Anilettan, Jinoj, Sanoop, Jayachechi, Pankajakshettan, Chandrashekarettan, Ambikachechi, Jaison Panikulangara, Robinettan, Shirleychechi and Unniyettan, Illias, Kusumam Teacher, Dr. Brahmaputhran, Dr. Narayanan, Vinithachechi, Yamini, Sureshettan, Sahadevettan, Ajayettan, Sumesh M., Baby John Saar, Prajil, Ramseena, Ajilal, and all those activists of various stripes who honored me by their willingness to engage me as an interlocutor. I am also grateful to those opponents of the campaigns I studied, such as police and factory employees, who were willing to speak with me despite the risks it often entailed, particularly Josettan, Naushadettan, Rajesh, and Davisettan. Finally, I wish to thank those who were neither activists nor their opponents, but who took enough interest in my research to share their opinions and allow me to accompany them during some small portion of their lives; Minichechi, Kenny, Radhakrishnettan, Muraliettan, Fr. John, Gokul, Pushpachechi, Ramakrishnettan, Paval, and Jithin are among these. During my approximately three years in Kerala, I was gifted with the hospitality, camaraderie, and conversation of far more people than I can mention here. I can only comfort myself by remembering that it is the return of friendship, not verbal declaration of thanks, that is considered the proper sign of gratitude in Kerala.

Special thanks go to my research assistant in India, pseudonymously called Ahmed here, who not only helped to collect much of what I analyze in this manuscript, but agreed from the very start to throw his own story into the pot. His contribution, as it turned out, became the salt in the soup. I am grateful that his story continues to be closely tied to mine.

I first developed many of the ideas in this book as a graduate student at the University of Michigan, where I benefited from the mentorship of Webb Keane, Matt Hull, Judy Irvine, Stuart Kirsch, Karen Staller, Lorraine Gutiérrez, Tom Fricke, Lawrence Root, John Tropman, Michael Spencer, Lee Schlesinger, Fred Vandenberghe, and David Akin. This project also benefited greatly from the guidance and mentorship of J. Devika at CDS, whose investment of time and sage advice in my formation as a scholar can never fully be repaid. Conversations with Shaji Varkey (University of Kerala), D.D. Nampoothiri (CREST), Vinod Krishnan (CREST), M. Gangadharan, M.G.S. Narayanan, and Nizar

Ahmed all improved my understanding of Malayali culture and politics and helped to shape the analyses presented here.

Over the years, many people have offered comments and edits on drafts of this book, guiding my hands as I pruned and reshaped it. My intellectual exploits over the last decade have largely been undertaken in the company of Charles Zuckerman, whose mark on this manuscript lies deeper, I am sure, than I can fully perceive. Duff Morton and Narges Bajoghli, likewise, had their fingers deep in the text at every stage—but more importantly, they helped me find my voice. Late in the process, Joseph Mathias's comments and edits on the entire manuscript helped me to fine tune that voice. I have also greatly benefited from dialogue and exchange with Dan Birchok, Meghanne Barker, Joseph Hellweg, Jeff Albanese, Jane Lynch, Ujin Kim, Mythri Prasad, Ellen Ambrosone, Navaneetha Mokkil, Amy Krings, Peter Railton, James Meador, Robin Zheng, Victor Kumar, Dylan Nelson, Michael Prentice, John Schwenkler, Nishita Trisal, John Doering-White, Amy Krings, Lisa Reyes Mason, Vincent Joos, Tam Perry, Naveeda Khan, Sahana Ghosh, Jason Cons, and Diane Mines. The Linguistic Anthropology Laboratory, the Evidence of Ethics in Action Interdisciplinary Workgroup, the International Society for Third Sector Research, Scholars Across Social Work and Anthropology, and a dissertation-to-book workshop with the American Institute of Indian Studies provided forums for critical discussion and feedback on portions of this manuscript. My anonymous peer reviewers each offered unique insights and suggestions that helped me improve the manuscript in both style and substance. Anas K. put his heart and his talent into the artwork for the cover. I am also grateful to the editors and staff at the University of California Press—to Stacy Eisenstark for taking the manuscript on, to Naja Pulliam Collins for believing in the project and guiding the manuscript through the revision process, to Catherine Osborne for spiffing it up, and to Emily Park, Teresa Iafolla, and all the others who are helping get it out into the world.

This work would never have come to fruition without the patient support of friends and family members. I wish to thank my parents—Mark Mathias, Barbara Mathias, and Sathya Reddy—for encouraging me even as I pursued a path unfamiliar to them. I found respite in the company of my siblings Mary Lapp, Graham Lapp, Anne Fleming, Steven Fleming, Chetthan Reddy, and Swathi Reddy. Walks with Ryan Bodanyi kept my mind inquisitive. The company of Samip Mallick, Gordon Simonett, Praveen Kache, Biju Edamana, Dieter Bouma, and Deena Thomas kept my spirit fed.

Finally, I wish to thank Geethanjali Mathias, whose baby smiles did more for my rapport in Kerala than all those years of preliminary fieldwork, and whose ten-year-old smiles are always the best reason to put my work down at the end of the day. But above all, I am grateful to Deepti Reddy. When things were going well, she reminded me to rejoice; when things were tough, she was my confidante and counselor. She accompanied me on visits to Kerala; she proofread drafts; she learned Malayalam and helped teach it to Geethanjali. Looking back on the path I have traveled, I see that the best part was that I traveled it all with her.

# Note on Translation and Orthography

Unless otherwise indicated in the text, all of the quoted speech presented has been translated from Malayalam to English by me, sometime with the help of a native speaker. My translations emphasize readability in English. Occasionally, I introduce a transliterated Malayalam word or phrase to discuss its meaning and usage. However, because I anticipate that most of my readers will not be able to "hear" Malayalam phonology in their minds, nor place isolated terms within a field of Malayalam-specific semantic relations, I avoid extensive presentation of transliterated speech throughout. When I do introduce key Malayalam terms, I generally offer an English term that can stand in for the Malayalam during extended passages, occasionally noting the Malayalam term parenthetically. When a key quoted term is a loanword from English, I indicate this with [Eng].

In transliterating Malayalam to Latin script, I have followed the system developed by Kunjan Pillai (1965) and shown in the table below. Symbols are listed in dictionary order (top to bottom, then left to right). I follow Asher and Kumari (1997) in transliteration of the central vowel "ə," represented primarily by the diacritic "ു്" in Malayalam script, which only occurs in word-final position. I have spelled words with commonly used romanizations (for example, *hartal*) in the usual way.

The term *Malayalis,* denoting speakers of Malayalam, is used throughout the text to refer to the people of Kerala.

TABLE 1. TRANSCRIPTION ORTHOGRAPHY TABLE

| Malayalam Alphabet and Latin Transliteration | | | |
|---|---|---|---|
| അ | a | ഠ് | ṭh |
| ആ | ā | ഡ് | ḍ |
| ഇ | i | ഢ് | ḍh |
| ഈ | ī | ണ് | ṇ |
| ഉ | u | ത് | t |
| ഊ | ū | ഥ് | th |
| ഋ | ṛ | ദ് | d |
| എ | e | ധ് | dh |
| ഏ | ē | ന് | n |
| ഐ | ai | പ് | p |
| ഒ | o | ഫ് | ph |
| ഓ | ō | ബ് | b |
| ഔ | au | ഭ് | bh |
| അം | am | മ് | m |
| അഃ | h | യ് | y |
| ക് | k | ര് | r |
| ഖ് | kh | ല് | l |
| ഗ് | g | വ് | v |
| ഘ് | gh | ശ് | ś |
| ങ് | ṅ | ഷ് | ṣ |
| ച് | c | സ് | s |
| ഛ് | ch | ഹ് | h |
| ജ് | j | ള് | ḷ |
| ഝ് | jh | ഴ് | l̤ |
| ഞ് | ñ | റ് | r̤ |
| ട് | ṭ | ററ് | t̲t̲ |

SOURCE: Adapted from Kunjan Pillai (1965).

INTRODUCTION

# Activist Lives

## WALKING THE EDGE OF ACTIVISM

As the anti-quarry march neared his home village, Ahmed began to hang further back, separating himself from the group and no longer joining in songs.[1] I walked with him and gradually we found ourselves alone, the rest of the marchers out of sight ahead. We were just entering a stretch of shops and restaurants, the main drag of a small town where Ahmed's neighbors and kin did much of their shopping. A man stopped us to ask where we were from, and Ahmed told him Palapuram—a town on the other side of the state, where he and I had been living with a family of environmental activists.

"I'm not going to tell anyone where I'm from," he explained when the man had moved on. "If I tell them my uncle's name, they'll ask me: 'Which Bashir? Quarry Bashir?'"

Over the previous six months, Ahmed and I had visited places where quarries had chewed forested mountains down to bald heaps—places where children coughed on fine, gray dust and parents took to the streets in protest, bearing torches and shouting slogans. Ahmed now strongly opposed the construction of megamalls, multiplex theaters, and other concrete extravagances. But his uncle's quarry business had supported him and his widowed mother all his life. His uncle had even paid for him to attend college—the same college that had introduced him to the notion of environmental justice and eventually led him into environmental activist circles. Ahmed was ready to march against quarries in other villages, but he was not sure if he could march at home.

"I'm scared," he explained. "Everyone here knows my uncle is the one who gave me my whole education. Then if I give him trouble in return. . ."

• • •

What is to be gained or lost by becoming an environmental activist? For the year we spent together, Ahmed was always on the edge of activism, weighing this question. He was my research assistant, hired to join me as I conducted anthropological fieldwork with environmental justice movements in Kerala, India. His qualification for the position was his recently acquired bachelor's in social work, but he had no prior experience with environmental activism. Ahmed was a devout Muslim from a relatively rural part of Kerala, and his year among activists had been a time of questioning, of taking risks, of waking to new consciousness. In this sense, it was an energizing, liberating experience. The pursuit of environmental justice gave his life clearer purpose and brought him into a community of people on a similar path. But at the same time, Ahmed struggled with his new friends' views on gender roles, marriage, and religion. When he was with them, he found it difficult to keep up his daily prayers and attend Friday services. Most of all, he feared the activists would discover his uncle's quarry business or that his relatives would discover his participation in anti-quarry campaigns. As the activists made quarries their central focus, Ahmed's environmentalism seemed to be driving him toward an inevitable clash with those most dear to him.

Such tensions between cause and community are the focus of this book. But not all activists in this book experience or respond to these tensions in the same way. For Faiza and Adarsh, the activist couple with whom Ahmed and I lived during our fieldwork together, leaving behind communities of kin, caste, and party had been part of learning to live for environmental justice. Like Ahmed, they experienced activism as socially risky, but embracing these risks was essential to their notion of a good life. The book also describes activists whose causes arise from their communities. For example, Sunitha joined an Action Council in order to put an end to pollution from a gelatin factory in her own village. Though there were those who disapproved of her activism, including some family members, her struggle for environmental justice grew out of her roles in her community, especially her roles as mother and neighbor.

Environmental justice movements in Kerala are structured by these two approaches to activist life: activism that opposes community to cause, and activism that combines them. We see the first approach in people like Faiza and Adarsh, for whom environmental justice signified a radical transformation of self and society. These radical environmental activists used a magazine, *Kēraḷīyam,* to elaborate and promote this broad vision for change.[2] For others, like Sunitha, activism begins from the direct impacts of environmental hazards on their daily lives. She and other local activists in her village of Manamur took the second approach, protesting pollution in defense of their own community. These two kinds of activists played mutually interdependent roles in environmental justice movements, and their stories sketch a range of possibilities for what an activist life can be.[3]

The quandaries that define the struggle for environmental justice in Kerala are rooted in dilemmas we all face. Each form of activist life speaks to tensions in the form of human life. Humans have the capacity to strive for a particular goal or purpose, to live for something. But they also come from particular communities, social locations, or circumstances; even as they live *for,* they always also live *from.* The radical activists' stories show what can happen when people attempt to hold these two aspects of life apart—to make their moral purposes independent of their community ties. The local Manamur activists' stories show the power and the limitations of attempting to combine the two synergistically—of living for the people and places one comes from. Each mode of activism brings tensions between *living for* and *living from,* cause and community, to the surface. Each casts these tensions in its own light, illuminating its own possibilities and giving rise to its own dilemmas. The insights to be gained are not only for environmental justice activists. Though many of us will never commit ourselves so fully to one path or another, these same tensions at times force all of us to choose.

Ahmed, falling ever further behind the rest of the marchers as he neared his village, felt this keenly. He was convinced that quarrying was destroying the environment, but he also did not want to betray his uncle. He knew, moreover, that many of the other marchers were estranged from their kin because of their activism. Should he stand by his convictions or be loyal to his community? What mattered more?

## LIVING FOR AN UNCOMMON CAUSE

Months before Ahmed faced his dilemma on the anti-quarry march, Faiza had spoken about a similar choice. Ahmed and I had joined her on a weekend-long nature program that she had organized for

FIGURE 1. Faiza leads college students through the forest during a nature camp.

some college students in a state-protected forest. All the official activities of the day were over, and we were gathered on a sort of veranda outside our sleeping quarters.

It was late March, and the night was cool, dark, and very quiet. I could later recall only two sounds, both steady and calm. One was the *shhh* of the river threading through boulders in the woods below. The other was the even, confident legato of Faiza's tuneful voice, talking about what it means to be an activist, a topic she had spoken about many times before. Two dozen students from an art and design college, all men, had arranged themselves around her on the cool concrete floor, legs crossed or splayed.[4] They had been noisy all day, chatting and joking as they picked up tourist trash in the nature preserve, drumming and singing as they rallied in the road, squealing and splashing in the cold river, calling out at the sight of a hornbill, a lion-tailed macaque, or a Malabar giant squirrel. But now they were listening.

Faiza never used the word "activist." Instead, from various angles, she presented a contrast between two ways of life. On the one hand was the life that the students had been pursuing until now: a life defined by the pursuit of high salaries, suitable marriage partners, and family approval. This is the usual life in Kerala, she said, but it is a life defined by greed; it feeds on the exploitation of both nature and people. On the other hand were the very sorts of activities the students had participated in that day. She spoke about limiting one's use of plastic, water, and energy; about protesting against environmentally destructive industries and development projects; and about finding new friends who share one's values.

"Perhaps it necessary to change humanity's whole way of life," she said. "Why can't such activities be the focus of a life?"

Thus, a choice became clear. But Faiza did not suggest it was an easy choice. She offered her own life as an example, describing how she had given up more lucrative career opportunities to work for a small non-profit, the River Protection Forum, and how she had become estranged from her mother when she, a Muslim, had married Adarsh, a Hindu. Salary, marriage, family approval—these things had been left behind. But these things, Faiza said, were not the things most worth living for.

"Life is not this thing that your families, societies, and teachers are always trying to scare you about," she told them. "I'm telling you this because you may have to face such decisions in the future. I hope my words may be helpful."

The art students were listening. Earlier in the day, a few had expressed that they, too, wanted such a life. For two days and one night, in a forest far from parents and teachers, they had been feeling out the activist life. And yet, even as Faiza talked, the gap between her life and theirs became apparent. In the middle of her speech, one-year-old Tara had begun to bawl for milk. As Faiza took her up and began to nurse, the young men shifted their gaze or turned away. She did it casually as she spoke, as if it were the most ordinary thing in the world, but Ahmed told me later that it was extremely unusual. He had found it impossible to look at her.

• • •

Activists are often defined by their willingness to be radically different, to take a stand for what they believe is right regardless of what anyone else thinks. As Martin Luther King, Jr. once noted, such people are sometimes labeled "extremists" and criticized for being inflexible, impertinent, or divisive. But they are also admired as moral trailblazers, people who have the courage to question the values of their peers and predecessors.[5] Some activists in Kerala see their own lives in this way. They distinguish their lives from those of "common people" (*sādhāraṇakkār*), whose values are shaped by the mores of their families, caste groups, political parties, or other communities.[6] Like Faiza, these activists tell of choosing a more liberated and rational way of life—a life guided not by the approval of others but by one's own vision for a better world.

If activist life is about putting one's principles over social pressures, cause over community, then environmentalism can seem to take activist ethics to its logical endpoint: an ethics without any community boundaries, even the boundaries of the human community. Environmentalism, some claim, represents the culmination of a progressive broadening of moral concern to ever-wider circles of belonging—from tribe, to nation, to humanity, to ecosystem.[7] In philosophy, this process is described as moral extension.[8] Extensionist environmental thinkers have called for a radical decentering of human concerns to make way for more expansive, ecocentric ethics.[9] Climate science has made this moral point a practical one: it is now clear that the survival of the human species will depend on broader attention to the needs of a global ecological system, not only to the desires and wants of the people residing on the planet today.[10]

In her conversation with college students at the nature camp, Faiza presented the choice between community and cause as fundamental to a way of life. For Faiza and Adarsh, fighting for environmental justice was not just a weekend activity. And while they both spent most of their weekday hours at environmentalist jobs—she at a small NGO focused on protecting riparian ecology and Adarsh at *Kēraḷīyam* environmental magazine—activism was not a career either. Often, the choice was a million little choices: organic or conventional, cotton or polyester, ayurvedic or allopathic, soft drinks or water, white sugar or brown. But it could also include singular, path-defining choices, like seeking out a spouse from a different caste or religion. It was building houses out of earth instead of concrete, or avoiding building any house at all. It was marching to the capital, attending seminars in the forest, kissing friends, nursing in public, and singing songs all night in the rain. In pursuit of social change, activists like Faiza and Adarsh explicitly lived counter to the values of those around them—that is, counter to the values of common people (*sādhāraṇakkār*) and, usually, the values of their own kin, neighbors, and peers.[11] Being an activist was being uncommon.

Notably, these activists did not see their concern for the environment as displacing concern for humans. They called environmental protests "people's protests" (*janakīya samaraṅṅaḷ*), and they fought not only for the environment (*paristhiti*) but also for the people (*janaṅṅaḷ*).[12] In practice, "the people" were identified with specific groups who had been impacted by development projects, environmental degradation, or industrial pollution. Activists like Faiza and Adarsh worked in solidarity with these groups to demand justice—thus, I refer to them not only as "radicals" but also as "solidarity activists."[13] But the aim of their solidarity work was not simply to win justice for specific impacted groups. Rather, they viewed group identities and norms as barriers to the pursuit of true justice for all, human and nonhuman alike.

The choice Faiza presented to the students was central to how she understood her own life, but it also captured how her life was seen by others. Radical environmental activists' moral aspirations set them apart—not only in their self-understanding but also in the assessment of those they called common. For many in Kerala, communities of kin, caste, religion, or political party were formative to their moral lives; they found their purposes in part through their roles in these groups. This was certainly true for Ahmed, whose notion of a good life was, above all, to be a good son, a good nephew, a good Muslim. It was also often true for those impacted by environmental conflicts, to whom activists like Faiza sought to offer their solidarity. From these perspectives, such single-minded commitment to a cause, even if admirable, could seem alien.

## FIGHTING FOR KITH AND KIN

Just outside the gate of the Manamur Gelatin Factory, jutting out from the base of the factory wall into the village's main road, lay a low platform of concrete, topped by a pavilion of bamboo poles and blue plastic tarps. This "protest tent" (*samarappantal*) served as the headquarters of the local campaign against the factory's pollution. On most days, it was a quiet place. As a rule, it was always occupied; an empty protest tent was widely understood as a sign that a campaign had lapsed or failed. Yet, by day, the tent was rarely very full. On any given afternoon, I could expect to find a few men there—usually the same few men, slouching low in molded plastic chairs and conversing intermittently, patiently, while dozens of colorful flags and posters proclaimed outrage on their behalf. But as dusk came on, the tent would fill and overflow. More men would arrive, talking loudly and standing in the street. Women would arrive as well, intermingling with their husbands, but also taking their own half of the platform, sharing their own conversations and jokes. And with the women came children, from toddlers to

FIGURE 2. Families gather in the protest tent under the Manamur gelatin factory's floodlights.

teens, who sprinted and swatted and squealed and sprawled and made of protest their own games, with the tent as their playhouse.

For those who gathered in the protest tent in Manamur, environmental activism was done with family and neighbors. And it was also done *for* them. As Sunitha, one of the most regular visitors to the tent, told me, "We mothers joined the Manamur Action Council in order to save our village (*nāṭ*)." Others came to the campaign as fathers, sons, or daughters—or simply as "locals" (*nāṭṭukār*), as those who belonged to Manamur.[14] Their activism was not opposed to group belonging. On the contrary, it was predicated on being a member of certain communities and not others.

The activism of Sunitha and other Manamur locals offers a distinct vantage point for exploring tensions between community and cause. Participants in the Manamur Action Council collaborated closely with activists, like Faiza, who saw environmental justice as an alternative way of life. They sought these activists' advice and assistance, learned from their expertise, and hosted them in their homes. They also took up the radical environmental activists' rhetoric, including universalizing abstractions like "nature" and "the people." Nonetheless, locals in Manamur did not take themselves to be fighting for radical social change, nor did they believe their activism conflicted with their existing group identities. For Sunitha, there was no reason that pursuing environmental justice necessarily meant estrangement from kin, caste, religion, or party—let alone neighborhood or village. She fought for nature by fighting pollution in her own well and along her village's stretch of river. When she demanded justice for the people, she was demanding justice for *her* people.

Just as activists like Faiza distinguished themselves from Action Council participants, calling them both "the people" (*janaṅṅaḷ*) and, at times, "common people" (*sādhāraṇakkār*), so Action Council members explicitly marked the differences between their activism and that of those they called the "environmental activists" (*paristhitika pravarttakar*). They acknowledged that the scope of their own activism was narrow by comparison, repeating frequently that their only goal was to stop pollution from the factory. They did not, with a few exceptions, stop drinking Coca-Cola or using Western medicine. They did not usually marry outside of religion or caste, nor take low-paying jobs in environmental NGOs. When one young farmer did start to talk about going organic, others teased him relentlessly. Being "the people" was a form of environmental activism, but it was different from becoming an "environmental activist." It did not require the kind of life choice that Faiza presented to the art students. One could be "the people" without giving up the life of the "common people."

The contrast between Faiza and Sunitha, or between those who fight in solidarity and those who fight for their own, is not unique to Kerala activists. It is reflected in tensions between environmental justice movements and "expanding circle" varieties of environmentalism. The concept of environmental justice was first explicitly put forward by activists in the United States in the 1980s and 1990s, many of them people of color, who fought against the disproportionate siting of power plants, incinerators, heavy industry, and expressways in their neighborhoods.[15] Like Sunitha, they grounded their environmentalism in their struggles to protect their own people. From the beginning, so-called mainstream US environmental activists, who were predominantly white and mainly concerned with preserving wilderness from human meddling, marginalized environmental justice movements as not truly environmentalist.[16] In part, they argued that environmental justice movements were too anthropocentric and thus did not fulfill the environmentalist vision for extending moral concern to all beings. On the other hand, proponents of environmental justice challenged mainstream environmentalists for ignoring social inequality and, at times, promoting racism.[17] Against the charge of anthropocentrism, they pointed to the antihuman tendencies in movements that focus solely on protecting nature from humans, without regard for social justice.[18]

Today, it may seem that proponents of environmental justice have won this debate decisively. The concept of environmental justice has now spread well beyond its US origins and is used to describe diverse movements around the globe that challenge social inequity in environmental matters.[19] Meanwhile, many mainstream environmentalist organizations, such as the Sierra Club and Greenpeace, have incorporated the principles and rhetoric of environmental justice into their work.[20] Yet the widespread acceptance of the environmental justice concept does not resolve the underlying conflicts because, as this book shows, environmental justice activism often straddles both sides of the debate.

While environmental justice movements in Kerala are organized by specific affected communities to fight for their own welfare, they are also collaborations between these communities and a range of "outside" actors, including activists committed to more expansive environmentalist agendas. Faiza, Adarsh, and others like them were active supporters of the protestors in Manamur. Faiza, who has an excellent singing voice, helped put out a CD of protest music on behalf of the campaign. Adarsh, as editor of *Kēraḷīyam,* covered the campaign extensively. Others marched, made posters, gave strategic advice, and generally participated in every aspect of the campaign. The stakes in protesting the factory's pollution were different for the radical environmentalists, but they were no less active than locals in the fight against the factory. Thus, tensions between fighting for all beings and fighting for one's own community were, in practice, integral to the collaborative structure of environmental justice movements.[21]

Locals also felt these tensions in their own lives, pulling them from either side. Fighting for one's own people could involve difficult choices not unlike those faced by radical environmentalists. For Sunitha, it meant giving less attention to her children's education and, as a result, watching them struggle in school. For others, it meant forgoing a day's wages in order to sit in the tent, blockade the factory gate, or join a march through neighboring towns. It meant arrests, court dates, lawyers' fees, and fines. And it also meant, at times, being willing to incur social risks similar to those experienced by Faiza and Adarsh. By the time Sunitha joined the Action Council, her husband was already a regular at the protest tent—now she and her daughters joined him. But as a result, she was no longer on speaking terms with some of her kin, one of whom was a factory employee. Others talked of losing friends over their activism or of being called traitors by neighbors and family members. Shop owners and auto-rickshaw drivers who supported the campaign worried about losing business.

But the stakes in these dilemmas were different for local activists in Manamur than for activists like Faiza and Adarsh. Manamur residents had not sought to become something other than "common people," nor did they see themselves as such. Their activism marked them as different from those around them, but they did not understand their activism as a search for an alternative life. Rather, through their activism, they fought to have ordinary lives. From this perspective, conflicts between community and cause are more of a bane than a banner, but they are nonetheless integral to life as an activist.

## *LIVING FOR* AND *LIVING FROM*

Faiza and Sunitha lead very different lives. Many of the values that Faiza rejects—commonly held notions of what makes a good career, a good marriage, or a good family life—are the values that define what Sunitha is fighting to protect. But this does not mean that one woman is more fully committed or passionate than the other. Both women's lives are highly focused on the pursuit of a moral purpose, a cause. Their differences speak to the range of possibilities for what it can mean to live for environmental justice—or, indeed, to orient one's life to any moral purpose at all.

Of course, living for a purpose is not unique to activists; it has often been seen as basic to what makes us human. The early sociologist Max Weber made this point forcefully, arguing that we cannot understand or interpret any human act without first understanding the purposes toward which that act is oriented.[22] Today, the notion that humans have purposes is intrinsic to ethnographic studies like this one, which seek to describe human lives as they are experienced and made meaningful by those who live them.[23]

But purposes do not appear from nowhere. The purposes we pursue are, to a large extent, about the communities, social locations, or circumstances we come from.[24] The core anthropological idea of culture is one way of talking about this social locatedness of human values and aims.[25] Thus, the differences between Faiza's and Sunitha's lives can be traced in part to differences in whom, what, or where they are from. This is already apparent for Sunitha, who frames her own motives in this way. But one could also narrate Faiza's activism like this: her father was a schoolteacher with Communist leanings, and some say she gets her nonconformist streak from him. Likewise, Ahmed's budding activism, which he feared might anger his uncle, could also be traced to the social work education that his uncle funded.

This is not the only way that stories about *living for* and *living from* can be told. Against an early tendency of anthropologists to reduce moral purposes to social conventions and group pressures,[26] recent anthropological research on ethics has focused on processes of critical reflection and self-cultivation.[27] Some have argued that ethics presupposes freedom, noting that even conformity to religious traditions and group norms requires active participation on the part of individuals.[28] For example, in describing the self-cultivation of piety among Muslim women in Egypt, Saba Mahmood argues that even submission to religious authority should, insofar as it helps these women to realize a desire for spiritual growth, be understood as an expression of freedom.[29] Similarly, if Faiza's life-defining choices can be traced to her father's influence, Sunitha's activism, which she attributes to her role as a mother, can be redescribed as a personal choice. Not every mother in Manamur joins the Action Council. When Sunitha sets out for the protest tent each night, she also sets herself apart from most of the other mothers in her family and neighborhood.

Even as anthropologists studying ethics have emphasized freedom, they have also retained a commitment to the idea that ethics is integral to social life. They have sought to capture this with notions like the social embeddedness and inter-relationality of ethics.[30] Yet these efforts to strike a balance can feel unsatisfying—in part because many anthropologists insist on distinguishing "ethics," as a domain of freedom and critical reflection, from the socially imposed rules of "morality." As others have suggested, the opposed terms used to make this distinction—such as deliberation and habit, practice and code, or agency and coercion—are often inextricably tied up with one another in actual social life.[31] Such dichotomies make it difficult to hold both *living for* and *living from* in view at once.[32]

How should we understand the relationship between these two aspects of being human? Recent emphasis in anthropology on activism makes this question particularly pressing. On the one hand, the lives of activists can seem to represent the polar extreme of *living for* rather than *living from;* they set out to break the norms of their given communities, and even to revise those norms to fit their own ideals. This theme is strong in studies of (and with) activists.[33] Yet accompanying this is an emphasis on how multiple identities, such as race and gender, intersect to shape activists' (including activist anthropologists') experiences, perspectives, ideals, and aims. Thus, in her introduction to *Decolonizing Anthropology,* Faye Harrison argues that the path toward an "anthropology for liberation" should begin by giving prominence to the work of "scholars belonging to neglected, peripheralized, or erased traditions that have long confronted and challenged colonial and neocolonial structures of power and economic relations."[34] Both emphases are also strong in environmental justice activism, which combines the expansive ideals of environmentalism with an insistence on prioritizing the agendas of oppressed communities.

This is not only a theoretical problem for the field of anthropology, but also a personal challenge for me and many other anthropologists. In recent years many anthropologists, including myself, have felt an urgent need to make their work politically impactful—to not only understand social life, but to transform it. When I started graduate school, I helped to start a new student group called Ethnography as Activism, a move that was received with some ambivalence by faculty. But today there is a growing consensus that anthropologists should be politically active, often by taking up the causes of those they study.[35] At the same time, feminist and anti-racist scholars have, for decades, challenged anthropologists to account for how their social locations shape their research, and these calls have also greatly influenced the field.[36] In some ways, this shift goes hand-in-hand with calls for a more activist anthropology; taking a position also means accounting for one's positionality. Yet in my own work, this emphasis on positionality clashed with my aspirations to be an activist anthropologist. As a white American, what was my proper place in the politics of those I studied in Kerala? Should I take up the cause of those who most shared my views? Or should I hold back, given that my views are those of an outsider?

The point of this book is not to solve such problems. In the pages that follow, I do not argue for anthropological emphasis on *living for* over *living from* nor vice versa. I do not side with those anthropologists who emphasize freedom and critical reflection in ethics, nor with those who insist on the importance of social norms and community bonds. While I tell stories of activist self-cultivation, norm-breaking, and consciousness-raising as well as stories of community belonging and social pressures to conform, I do not present any of these as *the* story of environmental justice activism in Kerala, let alone of activism more generally. In this sense, this book does not answer the question of how we should strike a balance between *living for* and *living from.*

Instead, this book explores this question as it emerges in the practical puzzles faced by people like Faiza and Sunitha, people living for a cause. In doing so, I take inspiration from the philosopher Ludwig Wittgenstein's suggestion that the answers to ethical questions can only be approached via forms of life (*Lebensformen*).[37] Wittgenstein developed the notion of "form of life" as he was grappling with some of the most fundamental questions of philosophy—questions about the boundaries between the self and the world, the difference between sense and nonsense, and the meaning of moral statements about right and wrong. Ultimately, he concluded that the highly abstract, either-or terms in which philosophers had posed such questions (such as opposing idealism and realism) could lead to logical confusion and dead ends. Instead, he recommended we approach such questions by exploring how various possibilities for structuring human life may pose and answer these questions differently. He stressed that ethical questions especially—questions about how one should live—are of this sort.[38] Only by living, by observing the lives of others, and by reflecting on the possibilities held out by these lives, can we work out answers to the best way to live.

The question of the relationship between *living for* and *living from*, cause and community, is this kind of question. It can only be answered by weighing the promise and limitations of different possibilities for living. The structure of environmental justice activism in Kerala lends itself to this task by offering two broad, contrastive approaches to activist life: that of the radical environmental activists, who take cause and community to be antagonistic and emphasize the former over the latter, and that of the so-called local activists, who take cause and community to be synergistic and make a cause of protecting their own communities. Not that any of the lives described here conform neatly to this dichotomy. Rather, just as the life choice Faiza presented to the art students was actually, as she lived it, made up of countless choices, small and large, so also each form of life opens up countless dilemmas and uncertainties—a form is not a formula.[39] The book feels out the contours of possibility via the paths of particular lives, each of which reaches for different aims and bumps up against different challenges.

Through these stories, it attempts to give the reader a sense of the stakes in various forms of activist life. It also, more tentatively, gropes toward a better sense of the stakes in the human form of life—that is, in the fundamental tensions we all face between purposeful living and community belonging.

Ultimately, like the question of *living for* and *living from* more generally, the aims of this book are not only scholarly, but also personal. On these pages, I have sought to work through puzzles that I have encountered in my own life, including in my life among Kerala's environmental activists. And it is my hope that, in light of the lives described here, readers can gain insight into the possible forms their own lives may take.

## STUDYING ENVIRONMENTAL JUSTICE IN KERALA

I came to Kerala looking for activism, and it was never hard to find. Partly, this was because of my own background. Beginning in college, I had been involved in solidarity activism with environmental justice movements in other parts of India.[40] My activist friends introduced me to environmental justice activists in Kerala, who were waiting to meet me when I first arrived. Even without this introduction, however, activism would have been easy to find, because Kerala is a place where activism thrives.

A long, lean state bordered by the Arabian Sea to the east and the Western Ghats mountain range to the west, Kerala is probably best known as a tourist destination—as a place of beaches and broad, green vistas; a honeymoon destination for newlywed Indians and a sun-drenched vacation spot for cold Europeans. Both in Department of Tourism brochures and on the lips of many Malayalis—the most common term for the people of Kerala—this is *God's Own Country*™.[41] But the state is nearly as famous for its protests as its resorts. Word of this reputation reached my ears long before I ever visited. Listening to my activist friends from other Indian states, Kerala grew in my imagination as a kind of social movement paradise—a place where peasants had (once upon a time) wrested the land away from the rich, and, more recently, villagers had shut down a polluting Coca-Cola plant.[42]

Thus far, I have presented Faiza and Sunitha's stories as windows onto the lives of environmental justice activists, pointing to resonances with broad themes in environmental movements around the globe. But the stories in this book are also, more specifically, stories about life in

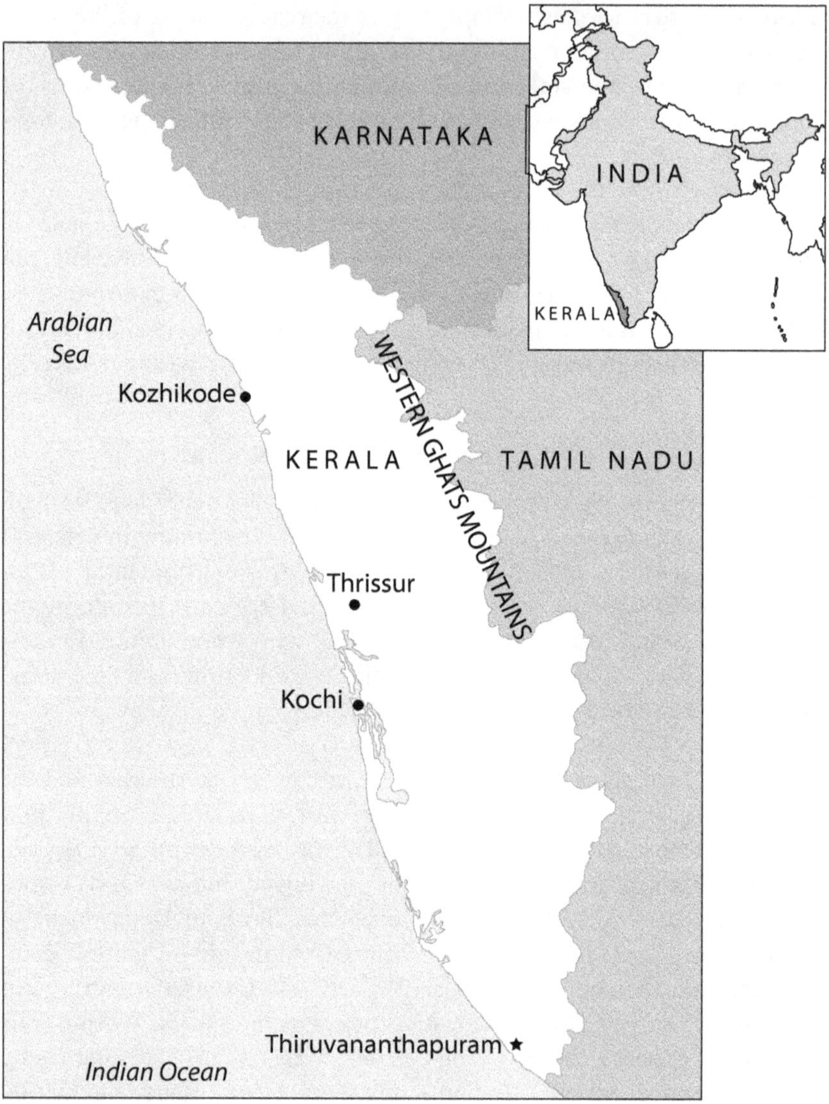


MAP 1. Map of Kerala. The pseudonymous village of Manamur is in central Kerala, between the cities of Thrissur and Kochi.

Kerala. The book's questions about *living for* and *living from* speak to alternative ways of seeing the state's history of strong social movements as well as contemporary public debate about the benefits and drawbacks of pervasive social protest. In studying environmental justice movements, the challenge was to tell both of these stories together.

*A Place of Many Causes*

My imagined Kerala was not a mere myth; social scientists and historians have long documented the unusually high activity and successes of Kerala's social movements. Prominent among these are the caste reform movements of the early twentieth century, in which oppressed castes fought to overcome some of the most extreme forms of caste discrimination in India.[43] Protest against British colonialism, including participation in the national independence movement, is also important to Kerala's social movement history.[44] Kerala's Communist movement grew out of both of these struggles, becoming the state's most famous movement as well as that most studied by scholars.[45] In the early twentieth century, Communist activists successfully translated a broad range of protests, especially against caste inequality, into a class-based struggle.[46] In 1957, when the state held its first elections, the Communist Party won a sweeping victory. Ever since, the Communists have been one of two dominant forces in Kerala's parliamentary system, the other being the Congress Party, the party of the national independence movement.[47]

Kerala's social movement history has often been linked to its development achievements—another major topic of social science scholarship. Beginning in the 1970s, Kerala became known for high levels of "human development" relative to its economic growth. Despite being among India's poorer states at the time, it was comparable to the United States and Europe in indicators such as literacy rates, infant mortality, and life expectancy.[48] The so-called Kerala Model of development presented an enticing puzzle to economists in the 1980s and 1990s because it seemed to contradict dominant theories about development, which see economic growth as a means of improving human well-being. Many scholars have seen Kerala's social movement history as an explanation for its unusual development pattern.[49] Under the rubric of "public action," they have described Kerala as a place where an organized, newspaper-reading, and politically savvy populace has held the government to account, successfully demanding policy that contributes to widespread social welfare.[50]

Within this broad story about social movements, democracy, and development, the central protagonist has usually been the Communist movement.[51] Prior to independence, the emerging Communist Party helped to cultivate a culture of reading and political participation centered on class struggle.[52] Later, building on their electoral successes, Communist parties and their eventual allies in the Left Democratic Front

(LDF) coalition passed and implemented extensive land reform, education, and social welfare policies[53] Moreover, even when not in power, the Communists have arguably so shaped the political culture that their political opponents have, at times, "continued and even extended social-welfare policies initiated by Left Front governments."[54] And LDF policy influence has, in turn, been credited with encouraging democratic participation and bolstering the power of social movements. In the 1990s, as national economic liberalization policies put pressure on states to relax labor laws and cut social welfare programs, the LDF-led government launched a People's Planning Campaign that distributed much of the state's budget to local governments.[55] Some scholars argue that this represents a new phase of the Kerala Model, in which grassroots public action and development are even more synergistically linked.[56]

Some have challenged this success story, pointing to those marginalized by Kerala's social movements or those left behind by the state's approach to development. Feminist scholars have shown that notions of progress, modernity, and human development, even as they seem to promote gender equality in some forms, have also introduced new norms of feminine domesticity that cast women as "weak and dependent."[57] J. Devika, whose historical work has critiqued the Kerala Model story from several angles, argues that caste inequality has only been reconfigured, rather than reduced, by so-called progress.[58] Others have pointed to how LDF policies have failed to prioritize Kerala's lowest castes and indigenous communities.[59] For example, while the land reform policies of the 1970s dismantled feudal land holdings and redistributed land to peasants and tenant farmers, it largely left out landless workers, the majority of whom were from the most oppressed castes.[60] Such arguments resonate with critical scholarship on development globally, which points to how the drive for progress and growth can perpetuate inequality. This raises the question of whether Kerala's development story is such an anomaly after all.[61]

Nonetheless, the notion of Kerala as a place of protests and progress—a place where activism has so permeated the culture that real change is possible—still fits with the self-image of many Malayalis today.[62] As I began my research, I found that Malayalis were quick to introduce their state as a place where caste names are no longer used, where colleges and hospitals are numerous, and where those begging at intersections are all migrants from other states. Among the college-educated, the words "Kerala Model" and "100% literacy" were as often part of these descriptions as *God's Own Country*™. And from the window of my hotel room, not far from the many white columns of the Secretariat building, the hub of state government, I could hear the din of chanted slogans and watch the colorful banners with their lovely, curling script, illegible to me at the time, passing in waves down the protest-packed street.

Looking back on these first impressions of Kerala now, I see the seeds for the main questions and concerns of this book. Early on, I wondered about the values that seemed implicit in Kerala's political culture. Why was there such broad support for rations, subsidized healthcare, and other social welfare policy here? What made people take such pride in widespread literacy? In hindsight, I see that I wondered how Kerala had become a place where a common purpose seemed to have transcended concern for self, family, or one's own community.[63] Though I did not have these words for it at the time, I can now say that I saw Kerala, and its Kerala Model, as a model of *living for*.

### *A Place of Many Communities*

But as I spent more time in Kerala, I had to question this view—not simply because it was overly romantic, but because this was not the main story that most Malayalis, in the twenty-first century, told me about protest movements in Kerala. Once I stepped out of activist circles, most people talked about protests as a problem rather than a point of pride. They eulogized the protest movements, or *samaraṅṅaḷ*,[64] of the past, and they acknowledged that protests are still pervasive today. But they said there were far too many protests. By and large, today's protest movements were not seen as the driving force behind Kerala's development achievements, but as precisely what was holding Kerala back.[65] People criticized marches, strikes, and rallies in the streets for driving away foreign investment and keeping busy people from getting to and from work. In an oft-repurposed joke, they complained that *God's Own Country*™ was being ruined by the "devil's own people." Rather than pointing to protests as indications of widespread solidarity or concern for others, they described them as expressions of greed and self-interest. They did not see struggles for justice; they only saw people blocking the road to further their own partisan agendas.

This criticism of protest movements presented a paradox: if people were so irritated with protests, why were there so many of them? Were they simply, as some suggested, stocked with people paid by special interests to march? Perhaps. But Adarsh, from his vantage point as editor of *Kēraḷīyam*, offered me a different explanation that others later affirmed. He said that, while nearly everyone is opposed to protests (*samaraṅṅaḷ*) in principle, every critic is also a member of a trade union, a party, a caste or religious organization, a business association, or any of Kerala's countless other groups and organizations, all of which from time to time take to the streets as well. People complain about protests in general, but they show up to protest on behalf of their own groups.[66]

Through Adarsh's account, I came to see Kerala's protest culture in a different light—one in which pervasive protest is an expression of *living from*, a sign of enthusiasm for advancing the interests and aims of one's own community.[67] This is clearly how protests are seen by their critics. But it also speaks to a basic local reality. Kerala is a place where people are organized; they belong to organizations. One need not go to the Secretariat to see this. On any neighborhood street, one can see these organizations' posters plastered on property walls, their signs planted at intersections, their insignia painted on the asphalt. Many of these organizations—especially the caste-based organizations and the political parties—are the products of Kerala's social movements. Yet they are also, today, groups into which one is born. This is true even of the communists: their politicians still often speak in an idiom of class struggle, a category that aspires to transcend community ties, yet today Kerala's communists are also just members of another party, another group identity—there are Congress families and Communist families.[68]

Even as I adjusted to this view of living-from as central to protest culture in Kerala, however, a key exception came into focus—an exception that included environmental justice protests. These belonged to the category of "people's protests" that were not affiliated with any party or organized group. As such, they ironically attracted far fewer people to their ranks than other protests that were not "of the people." Yet, by the same token, they also aroused less irritation. Over time, I learned that describing my research topic simply as "protests" (*samaraṅṅaḷ*) was likely to elicit annoyance or even disapproval. But if I specified that I was studying "people's protests" (*janakīya samaraṅṅaḷ*), I usually got a more positive response. Because issues such as a polluting factory or displacement by a new airport project were not prioritized by any major party or identity group, they were eligible for public sympathy. Many Malayalis described people's protests as the true heirs to the hallowed social movements of the past—especially the caste-reform movements and the independence movement. Compared to those movements, they were relatively uncommon causes, in the sense that very few people took them up. Yet, insofar as they were not partisan causes, they might also be seen as common causes—as protests on behalf of, rather than at the expense of, the people.

In part, the exceptional status of environmental justice movements in Kerala is deliberately produced by activists themselves. Activists like Faiza and Adarsh, who seek an alternative to the politics of group affiliations, work hard to craft people's protests as a domain that transcends such affiliations. But environmental justice movements also lend themselves to this project because, in the context of global capitalism and Kerala's ongoing push for development, no major parties have taken up their flag. At least, no party does so consistently. As one long-time environmental justice activist put it, people's protests are like "a stick" (*vaṭi*) that parties take up when they are not in the ruling coalition, self-righteously battering the rulers in the name of the people. But when these same parties are elected into government, they quickly throw the stick aside.[69]

As the metaphor of the stick makes clear, environmental justice activism—and, more generally, the category of people's protest—cannot easily be disentangled from the contending communities and organizations that dominate Kerala's politics. But the special status of people's protests nonetheless speaks to deep cultural currents in Kerala's social movement tradition, raising questions about how this tradition should be understood. What makes social movements so vibrant in Kerala? Is it the ability of people to transcend community boundaries, to shrug off a caste system and embrace class consciousness? Or is it the very organization of people into contending groups, ready to take to the streets for their own interests?

### *Subverting the Either-Or*

Such either-or questions must be handled with care; they echo problematic dichotomies long prevalent both in public discourse within South Asia and in scholarship about the region. In mass media, these dichotomies surface in discourse about South Asian "communalism," a colonial-era term for religious sectarianism that has long been used to mark a gap between Eastern politics and Western ideals of rationality and social detachment in public life.[70] The same dichotomies creep into the study of South Asian politics when these are measured against notions of an ideal "civil society" or "public sphere," in which individual citizens can engage in rational debate without interference from social affinities or group loyalties.[71] Similarly, the anthropologist Louis Dumont juxtaposed a cultural emphasis on social roles and hierarchy in India with the individualism and egalitarianism supposedly prevalent in Western society—an argument in line with early anthropological comparisons between the community-based morality of "primitive," non-Western cultures and an emphasis on freedom and universalism in their own "modern" societies.[72]

Subsequent scholarship in anthropology and history has challenged these dichotomies again and again. In some cases, anthropologists have undermined them by demonstrating the importance of culturally specific modes of individualism, expressions of freedom, and universalist moral projects to South Asian politics.[73] Studies of activists have been important to this work, and this book builds on these interventions. In other cases, anthropologists have underlined the continued importance of community identities to politics and morality in the region while also showing how South Asian modes of community affiliation confound Western categories.[74] Historians have likewise sought to show how the sectarian divisions today called "communalism" are not essentially South Asian but, rather, were historically produced by colonial governance.[75] Nonetheless, as Hindu nationalism has come to dominate Indian politics, global alarm over communalism has intensified. It is against this background that Kerala, where the Hindu nationalist movement remains relatively weak and this alarm feels less intense, can appear exceptional and enigmatic.[76] By the same logic, people's protests, when juxtaposed with the protest politics of parties and organizations, can appear to be an exception within the exception—and the lives of activists like Faiza can appear most exceptional of all.

The contrast between *living for* and *living from* may seem to parallel, or even reassert, the dichotomies that give rise to such appearances of exceptionality. But the purpose of this new contrast is to undermine the others. These other contrasts have been used to sort people (or nations, or civilizations) into communal and universalist buckets or to place them on gradients between one pole and the other. But *living for* and *living from* are aspects of all human lives—and although they are often in tension, they are not inherently opposed. They can be used to define different moral projects and to navigate different paths for one's life. But the relation is not one of opposition—north and south, or latitude and longitude—but rather of two forces that can, like wind and current in sailing, variously be in tension or aligned. Not two poles, but two pulls.

What differentiates Faiza from Sunitha is not that one woman's life is more *for* and the other's is more *from;* the difference is in their efforts to either separate or combine these aspects of their lives. And in each case, by closely examining how these projects of separating or combining are put into practice, we see the limits of difference between these projects, the common stakes.[77] Common to environmental justice activists in Kerala, yes, but also common in a more open-ended way. As Faiza and the radical environmental activists attempt to purify activist causes of community pressures, stretching the limits of moral extension, they remind us that all purposes are lived out in relationship with others. As Sunitha and the activists in Manamur seek to make their local community their cause, they bump up against contradictions that remind us of what is special about committing oneself to a cause. These are stories of alternative paths, but they are never so clearly divergent as the either/or contrasts that pervade applications of notions like modernity, civil society, or the public sphere to studies of South Asian politics. Even very different activist lives are nonetheless contiguous and comparable. In the same vein, I hope these stories can lead my readers to fruitfully reassess the tensions and dilemmas that arise in their own efforts to change their worlds.

### *Many Stories at Once*

By living alongside activists, I came to see that the story of Kerala's environmental justice movements cannot be well told either as a story of transcendent ideals nor as a case of competing communities. Instead, it had to be told in both ways at once. To do so, I divided my time roughly evenly between radical activists and local activists, hoping to achieve an equivalent depth of understanding about particular lives in either group. To study the radicals, I made a field site of *Kēraḷīyam* environmental magazine, which covered people's protests and helped to coordinate solidarity efforts. To study the locals, I focused on the campaign to shut down a gelatin factory in the village of Manamur. During the time of my study, this was one of the most prominent people's protests in Kerala. It was also the recipient, for several months, of intense collaboration and support from a Solidarity Committee formed by activists connected with *Kēraḷīyam*. In my fieldwork with either group, I followed the channels opened up by the relationships available to me—seeking not only to observe activists' lives but to enter into conversation with them about my observations and emerging analysis. Such conversations, I hoped, would challenge my own assumptions and give me insight into how they saw themselves and one another.

Even as I attempted to give each group of activists equal attention, however, my relationships were shaped by their differing approaches.

Early on, I felt less like an outsider among the radical environmental activists. Because they aspired to form relationships based on a common cause, rather than common origins, my own activism made a kind of belonging possible that I had not felt in other settings. For most Malayalis, most of the time, the legacy of British colonialism and the contemporary economy of European tourism fixed me squarely in the category of *sāyipp,* a colonial-era honorific that, in a postcolonial context, leans pejorative.[78] But the radical environmentalists tended, from the start, not to mark this aspect of my identity—not only by not calling me *sāyipp,* but by generally treating me neither with the unearned admiration nor the disapprobation that sometimes went along with being a white American in Kerala. I was still plainly different; my presence in Kerala was still predicated on privileges—especially, money and mobility—which were supported by the very systems of global inequality that people like Faiza and Adarsh were fighting against.[79] Nonetheless, these activists also had a long history of engaging with foreigners in this struggle; they were familiar with Thoreau, Leopold, Fukuoka, and, of course, Marx.[80] And many of them treated me like no one special—like a relatively young, inexperienced activist, a student in need of guidance. This opened the way to relationships based on shared values and desires, and even to some close friendships.[81]

Relationships with local activists in Manamur took longer to develop, in part because my own activism was less analogous to theirs. A child of public-school teachers from a midwestern college town, I had never had to breathe smog from factories, lose my home to dam construction, or even wonder about the lead content in my drinking water. I had fought for environmental justice in solidarity, but never as a local. Moreover, while there were a wide range of caste, class, and educational backgrounds represented among both the radical environmentalists and the local activists in Manamur, I generally found it more difficult to explain my research to the Manamur activists. The radicals practiced a kind a of activist inquiry that involved extensive reading of philosophy and social critique; discussions in the *Kēraḷīyam* office were not so different from discussions in my doctoral seminars back home. Conversation in the Manamur protest tent focused more on the doings of the factory and campaign strategy, often mixed fluidly into local gossip, and I had less to say about any of this. Even now, it is hard to know what Vijayan, Sunitha, and other local activists will make of this book.[82]

Paradoxically, as my relationships with activists in both groups deepened, I also felt the gaps between their position and mine more keenly. I felt myself to be more, rather than less, an outsider. This, too, had to do with the differences between my activism and theirs. For Faiza, Sunitha, and many others, environmental justice was a total way of life to a degree that my own activism had never approached. Nor was my fieldwork a form of activist research in the usual senses of the term.[83] Even as a participant observer, I was severely limited by my position as a noncitizen studying protests that often targeted the state.[84] My lack of participation was not, generally, a barrier to rapport—perhaps, in part, because both radical and local activists seemed to assume that I would have been more involved in activism it if were allowed. Yet, on a personal level, the situation was often uncomfortable. The more I understood the lives described in this book, the more my own work for environmental justice seemed limited, partial, or even half-hearted. The book's concern with belonging is, in part, shaped by my own struggle with whether and how I belonged among activists.

The limits of my activism, though uncomfortable, also opened up opportunities for relationships with nonactivists, and this proved fruitful in unexpected ways. While much of my time in Manamur was spent in the protest tent, I also got to know many who supported the factory or simply avoided the issue altogether. I saw how environmental activism impacted the lives of the workers at the gelatin factory and listened to their perspectives on the stench in the air and the slime in the water. Likewise, while living with Faiza and Adarsh, I spent much more time than either of them among our neighbors—and from them gained insight into how activists' lives were seen by self-identifying "common people" (*sādhāraṇakkār*). My main focus remained on the activists' stories, but my time among nonactivists helped me to appreciate what made these stories special and what they shared with more ordinary lives.

The most fruitful relationship afforded by the limits of my activism was my relationship with Ahmed. Late in our work together, I observed that Ahmed was struggling with belonging among activists as much as I. He too was neither fully an activist, like Faiza and Sunitha, nor entirely not an activist. Yet our positions were, in many ways, at opposite poles. Ahmed was from a Muslim family in a relatively rural part of Kerala. He had just completed his BSW, the first person in his family to attain a college degree. He was eager to learn more about regional social issues and, hopefully, go on to further studies, but environmental justice movements were entirely new to him. And while he talked of being inspired by activists in Manamur and at *Kēraḷīyam,* he was unsure about whether he could become one of them. If my environmental justice activism was mostly in my past, his was (possibly) in his future.

As such, Ahmed helped me to see activist life from the edge. Whereas activists' retrospective stories about their lives tended to have clear moral trajectories (from ignorance toward consciousness, from restriction toward freedom), it was not yet clear what direction Ahmed's story would take. For him, environmental activism was an emerging possibility that held both risk and promise, potential for loss and potential for gain. Walking with him at the edge of activism helped me to appreciate the stakes in the line between activist and "ordinary." It also made me keenly aware of how others I observed, whether they described themselves as activists or as ordinary people, also found themselves straddling that line.

## PLAN OF THE WORK

Near the end of our fieldwork together, Ahmed found out that he had been admitted to the master's in social work (MSW) program at Tata Institute of Social Sciences, the most prestigious social work program in India. It was something he had been working toward all year. In his application essay, he had written about his experiences studying Kerala activists and his newfound passion for environmental justice. And he saw the opportunity as a continuation of this journey, both as a researcher and as an activist. We were all proud of him. But on the night of his departure, as we reclined amid the greasy remains of a celebratory feast, Faiza and Adarsh also offered Ahmed some gentle criticism and advice.

This mode of conversation was unusual in our house. Though Adarsh and Faiza were a good ten years older than Ahmed—enough that they might have assumed the right to tell him directly what he ought to do—they generally sought to encourage change more indirectly, even when it came to asking him to help clean dishes and other mundane matters of sharing a home. But in this final conversation, their advice was explicit and challenging. They felt that his experiences living among activists ought to have changed him a great deal, but they were concerned that such changes had not happened. Faiza, in particular, observed that, despite having met people who had different ideas, beliefs, and lifestyles, Ahmed was still very much a "normal" person with a "neutral" approach to life. She was worried that his pursuit of an MSW might only exacerbate these qualities.

"These MSW graduates don't have any kind of awareness or sense of reality," she told him. "They see this 'social work' as something very separate [from their relations to family or their communities]. But you have seen a lot of things. You have directly come to know the pulse of people in a place like Manamur. All of that should be reflected in your personal life."

• • •

For Faiza and Adarsh, the pursuit of social change was not simply about getting a certain degree or undertaking a certain career; it was a total life. But what is this life? What does it require? These were the questions Ahmed weighed as he struggled to position himself among the people we studied. They are also questions to which Faiza, Adarsh, Sunitha, and other seasoned activists constantly returned. Faiza's criticism of Ahmed may make it seem that she believes she has all the answers, but environmental justice activism was fraught with ambivalence and uncertainty for her as well. For Adarsh and Faiza, their real answers were always still being worked out in the ways they lived their lives.

The book's structure reflects this open-endedness, this notion of activist life as an ongoing process. Chapter 1 explores how radical environmental activists sought to live in a way that could transcend their own identities and community ties. They traveled across Kerala to fight in solidarity with "the people," but the main work of living "for all" ultimately happened elsewhere—in offices and on beaches—where they gathered with one another. Chapter 2 turns to the local protestors in Manamur, who readily adopted the title of "the people" and welcomed support from the radical activists, but rejected the latter's vision for broader change and, instead, made local belonging central to environmental justice activism. The chapter describes how they sought to combine community and cause and asks why this ultimately led to conflicts with the radicals.

In chapters 3 and 4, I probe the limits of these two approaches to environmental justice activism, exploring how activists contended with contradictions and dilemmas posed by each approach. Chapter 3 describes how putting cause over community could strain the radical activists' relationships with one another, at times leading to social marginalization and even isolation. Chapter 4 examines the Manamur protestors' efforts to reach beyond their own circles of local belonging and gain the sympathy of audiences further afield, who would never smell the factory's pollution for themselves.

The book's conclusion returns to Ahmed's struggles at the edge of activism, asking what his story can tell us about activism as a special form of life. I consider Faiza's criticism of him as too "neutral" and "normal" and explore why, in her view, Ahmed fails to qualify as an activist. But I also delve more deeply into Ahmed's experience in order to reflect critically on what Faiza's framing highlights and obscures. Ahmed's predicament sheds light on how the pursuit of moral freedom can, itself, be another form of submission to group pressures. The path he ultimately chose offers an alternative view of what is to be gained or lost in living for a cause.

1

# Living for the People

## ARE WE NOT ONE?

A few weeks into the monsoon season, Sunny invited me to attend a "rain camp" (*maḷa kūṭṭāyma*). Sunny was the founder of *Kēraḷīyam* environmental magazine and, technically, its lead editor; many people called it simply "Sunny's magazine." But he often seemed less concerned with the publishing work itself than with the social activities surrounding it. These included public events such as lectures, conferences, and book releases, but Sunny was most fond of the more intimate, overnight gatherings called *kūṭṭāymakaḷ*—a term sometimes translated as "fellowship," but which environmental activists glossed, in this usage, with the English "camp." I had been to a song camp and a full moon camp, but this would be my first rain camp.

The main point of a rain camp is to sit out all night in the rain. Tonight, however, we began indoors, in a small beach house that one of Sunny's many friends had loaned him for the occasion. There were about twenty of us. We spread out on the living room floor, the ring of our bodies expanding as more came in, until every back was against a wall. We just barely fit. Then, Sunny called for a song.

Are we not one? (clap clap clap) *Nām onn allēē? (clap clap clap)*
Are we not one? (clap clap clap) *Nammaḷ onn allēē? (clap clap clap)*

This was a common song in this crowd—a song that I had heard again and again at marches, sit-ins, and other public events, including

FIGURE 3. Rain camp participants gather on the beach under a light rain and a nearly full moon.

my previous camps. Faiza once suggested to me that it was popular because, with its simple lyrics and slow cadence, it was the easiest song to remember. But I think it was also loved because it so insistently expressed a value that these environmental activists held dear.

| | |
|---|---|
| Are we not one? (clap clap clap) | *Nām onn allēē? (clap clap clap)* |
| Are we not one? (clap clap clap) | *Nammaḷ onn allēē? (clap clap clap)* |
| The keepers of ourselves (clap clap clap) | *Namukkuṭayōrummm? (clap clap clap)* |
| And the keepers of this soil (clap clap clap) | *Ī maṇṇinuṭayōrummm? (clap clap clap)* |
| Are we not? (clap clap clap) | *Nammaḷ allēēē? (clap clap clap)* |
| Are we not? (clap clap clap) | *Nammaḷ allēēē? (clap clap clap)* |

The repetition of a simple question—"Are we not one?"—drowned out the breaking waves. The lyrics never explicitly answered this question. Rather, with all voices singing in unison, all hands clapping in sync, we became the answer. According to some, Kerala's environmentalists had learned this song from indigenous activists in the rainforests of the state's western mountains—people fighting to sustain their own unity with one another and with the earth. The rain camp was meant to be part of a similar struggle. Sitting, singing, and getting soaked together, we would build community with one another. Declining to open umbrellas or seek shelter, we would bridge divisions between humanity and nature.

• • •

In the previous chapter, I described how Faiza sketched out a choice for students at a nature camp, a fork between two life paths. She and Adarsh had chosen their path, organizing their lives around the pursuit of environmental justice, and they chose it again many times each day. But living for environmental justice was not only a personal decision. Faiza and Adarsh's path may have distanced them from their families, but they were not setting out on their own. Rather, they were joining a long tradition of Malayalis who had set aside the values of their families, religions, and parties, and sought an alternative approach to life—not necessarily environmental justice activism, but an approach that similarly sought to transcend community boundaries. As they exited their communities of origin, they entered a sparse but lively network of fellow environmental justice activists that spanned the state.[1] By laying her own defining choice before the art students, Faiza was inviting them to join her in pursuing a life in which all people, and all of nature, could become one.

How can people make a life aimed at larger purposes, beyond the particular interests, traditions, and pressures of the communities into which they are born? For Kerala's radical environmental activists, living for environmental justice meant prioritizing values over relationships. The process began with departure—personal departures from their own communities, but also a more general distancing from the mainstream politics of group loyalties and interests. Yet this distancing was not a rejection of community life altogether. Rather, it was part of a larger effort to construct a new politics—an environmental politics of "the people," a category that transcended group distinctions. As with prior generations of activists, this effort centered on small-circulation magazines and the forms of social interaction they sustained. Through publishing and reading *Kēraḷīyam*, activists formatted a motley array of local protest movements into a politics without parties, credos, or manifestos. Soaking in the rain together, they made themselves the few who fight for all.

## LEAVING THE PARTY

On a level, grassy spot halfway up the side of a hill, a handful of eager young activists chattered as they unstacked red plastic chairs and arranged them in a circle. Then they hovered about as twelve men and women, older and quieter, settled into the chairs, greeting each other with pressed palms and polite nods. Further up the hill, khaki-clad police leaned on jeeps parked at the side of the road, watching and waiting. A circling TV cameraman tightened his orbit as Dhanya—a forty-something woman with a quick smile and a strong voice—took her own seat in the circle and called for the meeting to begin.

One by one, each of the twelve described noise, dust, broken hilltops, and felled trees. They recounted how they had formed Action Councils to resist, entering into relentless battles with quarry owners, politicians, police, and hired thugs. None had yet succeeded in shutting down a quarry. All were grateful for this chance to gather with others like themselves.

Dhanya had been working toward this meeting for months. She and other environmental activists had researched the quarry industry extensively and published articles on its evils. They had visited dozens of quarry campaigns, offering their solidarity and support. Over the past fifty days, they had walked the entire length of the Western Ghats mountain range in Kerala, singing songs about social and ecological harmony, sleeping on the concrete floors of school cafeterias, and holding council with every quarry opponent they could find. This was the same long march, called the "Dialogue Journey," that had passed through Ahmed's home village a month before. The journey itself had been a formative experience for many young people, including Ahmed, and some said this was its main aim: to inspire a new generation of environmental activists. But for Dhanya, this hillside, this gathering of the leaders of anti-quarry movements across Kerala, had always been the final destination. Now she listened patiently to their stories in the hope that the goal of all her work—a state-wide, grassroots coalition of anti-quarry campaigns—was finally being realized.

What moved Dhanya to hope for this? She and other radical environmentalists had brought the leaders of quarry protests to this hillside, but who or what had brought *her* here? By the time Dhanya took her chair in the circle of quarry protest leaders, she had long since organized her whole life around the struggle for environmental justice. She still maintained a career as a part-time lecturer in a local college,

FIGURE 4. Dialogue Journey participants carry banners and sing songs on the final day of the fifty-day journey.

teaching computer skills, and she was a single mother to two teenage sons. But, for the past six months, she had been a constant presence at meetings on Western Ghats environmental policy and trainings for the Dialogue Journey, even as she and a friend worked on a documentary on the impacts of quarrying. Before that, she had been the chief organizer of solidarity efforts on behalf of the protest movement against the Manamur gelatin factory, coupling this work with a documentary on water rights. Thus, during the Dialogue Journey, she was seen by younger participants, such as Ahmed, as an accomplished activist and leader.

Yet to many activists her own age, Dhanya was still relatively new on the scene. Her solidarity work in Manamur had made her a key figure in one of the hottest environmental conflicts in Kerala, but it had also been one of her first efforts at environmental justice activism. There, as on the hillside, she had found herself leading others in environmental protest. But much was still new to her, especially in comparison with activists like Sunny, who had been connected with environmental justice movements for decades. When the solidarity effort in Manamur eventually failed, more seasoned activists had reminded one another about her lack of experience.

### *Departure as a Life Choice*

If environmental justice was new to Dhanya, activism was not. In her own telling, her roles in the Manamur campaign and the quarry coalition were only the latest chapters in a much longer story of fighting for social justice—one that stretched back to her earliest years. She was the daughter of a prominent and very active member of the Communist Party of India (Marxist), or CPM, the largest of several Communist political parties in India.[2] As a child, her father's party activism had shaped her values and, in college, she had joined the Students' Federation of India (SFI), the student wing of the CPM. But shortly after college, as she started a family with another CPM activist, she began to read extensively on globalization and environmental issues. She grew dissatisfied with Communist positions on these issues, but she did not feel there was room within the Party to even discuss, let alone correct, its philosophical and policy failures. She remained sympathetic to Marxist ideology, but she felt Party membership—and, indeed, membership in any organization—would only inhibit the kind of open-ended reading, discussion, and problem-solving to which she was drawn.

"I saw that there are a lot of problems with traditional Leftist philosophies," she explained. "And I'm just trying to figure out how we can fix the philosophies that I used to think were right. But in an organization. . . . The way I see and understand things, there isn't any organization that works like that. Once you join, they make you do things their way. They won't let you just continuously discuss and make modifications. They won't accept that. So since I didn't want to do things that way, now I really don't have any organization. I belong to no organization."

In Kerala, activists' stories about getting into environmental justice activism often began with the Communist Party. More precisely, they began with departures from the Party. Adarsh, like Dhanya, grew up in a Party family. In recounting how he became Sunny's assistant editor at *Kēraḷīyam*, he began by recalling how, as a child, he had eagerly awaited the next issue of *Sovietland*, a glossy, full-color magazine published out of Moscow. He had enjoyed the little stories about the feats of Gorbachev, but he had even more greatly prized the magazine's pages as protective wrappers for his schoolbooks. Most children only used the lusterless, black-and-white newsprint of Malayalam dailies to wrap their books; his books marked him as the son of a Party member. Like Dhanya, he too joined SFI in college. He and Faiza both were active in their campus chapter, where they met. But also like Dhanya, they now looked back on their Communist activism only as a point of departure for their eventual rejection of party politics and entry into activism on behalf of nature and the people. As fondly as he remembered *Sovietland*, Adarsh now saw it only as Party propaganda.[3]

Others had similar stories. Not every journey into environmental justice activism began by exiting the CPM, but they nearly always began with *some* exit—some departure from the community ties of one's youth, usually from those communities into which one had been born. Some activists came to environmental justice from families aligned with other parties. Some, like Faiza, talked about communities defined by religion. This is how Adarsh and Faiza told the story of marrying one another. Both of their families had opposed the marriage, and they had celebrated their wedding with other activists, singing protest songs and eating natural foods in a forest preserve far from either of their family homes. Other activists told of refusing the career paths chosen by their families, or of becoming atheists. Some spoke of pain and bitter loss, of rejection.

Such separations were not always irrevocable. When Faiza's mother was struck with a terminal illness, she and Faiza reconciled to some extent. After her mother's death, her father began to visit more frequently. He even joined the Dialogue Journey for several days when it passed near his home. Nonetheless, even when relationships healed, stories of departure continued to frame activists' lives.

Activists made this framing salient, in part, by reinforcing past departures in their present interactions. While religious community, even if forsworn, still usually remained encoded in activists' names, caste was far more systematically pushed into the past. Activists stringently avoided caste markers in social interaction. Caste names were not used, and dietary markers were obscured by the adoption of environmentalist ethics of eating.[4] Though an activist would sometimes mention their caste background to me, particularly if they were from an oppressed caste, it was considered unseemly to ask about, and activists claimed to have little knowledge of one another's castes.[5] For example, when one couple shared with me that they were of a Dalit caste status, they asked me not to share this information with others in the group. They said they were not afraid that other activists would look down on them, but they also appreciated that no one ever tried to figure their caste out, and they believed no one knew. At least in this social circle, they felt they had been able to leave their caste behind, and they wanted to keep it that way.

In a different way, activists pushed social class into the past as well. When I first encountered *Kēraḷīyam* and the people connected with it, I took them to be roughly middle class. When I asked about their jobs (*jōli*), many of them talked about things like putting on magic shows or making toys from trash, and some just laughed at me for thinking this was a relevant question. In my experience, only financially secure people could think about work in this way. Yet over time, I saw how Francis's choice to make his living conducting magic shows and selling environmentalist books had been a source of considerable financial hardship and precarity for his family. On the other hand, I also saw how some had found economic opportunity in activism. Saleem, had gone from manual labor as a painter to (via his self-education as an activist and the relationships it engendered) having a seemingly more secure life as the leader of a statewide organic farmers' association.[6] More generally, activists' ambivalent attitudes toward money and *jōli*, and the life paths these attitudes entailed, made the usual criteria of class status difficult to apply.[7]

Taken together, these are not stories about leaving behind any one ideology, organization, or community. They cannot be easily classified as stories of middle-class activists or the stories of any particular caste group or stratum.[8] Nor were they simply stories of rejecting a certain set of values or beliefs in favor of another. Indeed, even if they regarded Kerala's Communist parties as lacking when it came to environmental concerns, Dhanya and many others still concurred with Marxist analyses on many points.[9] The stories they told were about leaving behind particular *kinds* of community: communities of kin, religion, caste, or party—the very communities that they understood to be most important to mainstream politics. And these stories were inextricably tied to the story of the unity and harmony that they sought.

This connection was palpable in Sunny's rain camp. Moments after we finished singing "Are we not one?," Sunny made a brief introductory speech, explaining why he had invited us and describing the tradition to which the camp belonged. He said that camps are meant to bring together "people who are different" (*vyatyastarāya āḷukaḷ*) for shared experiences and increased awareness. He told us that he had invited us because we were people who "can envision things differently, who can stand apart and see differently . . . who intervene differently." Here, he said, such people can "experience relief and make bonds."

As we did a round of introductions, I heard this point echoed again and again—what we shared were our differences from the world beyond. Peter earned his income as a medical technician, but he really saw himself as a farmer; in his free time, he was developing new organic agricultural techniques that he hoped would someday be profitable enough to support his family. Fr. Sebastian was a Catholic priest, but, he added quickly, he was not like the priests we had met; he had no parish and spent all of his time traveling around supporting people's protests. Preethi had been a teacher in a government school until a few weeks before, but now she had resigned and was collaborating with Saleem to found an environmentalist commune. Each of these stories fit the speaker into a common category—technician, priest, teacher—but only in order to stress some difference from others in that category: not your usual technician, not like priests you've met, not a teacher anymore. What the rain camp participants had in common were their uncommon lives.

### *Departure as a Tradition*

Even as becoming an environmental justice activist meant breaking ties and becoming different, it was not necessarily a lonely process. Moreover, activists did not take themselves to be the first to set out on such journeys; they imagined their personal exodus narratives as threads in a larger history, dating back to the beginnings of representative democracy in Kerala. Their environmental justice politics of "the people" was only the latest iteration of this tradition, which they described as an alternative (*badal*) mode of politics.[10] Likewise, they described *Kēraḷīyam* and the small magazine publishing tradition to which it belonged as alternative or parallel (*samāntara*) media. Before Dhanya left the Party in search of a different mode of politics, and before she found what she sought in protests of gelatin factories and quarries, there were others who made similar exits but found different answers. Before Sunny founded *Kēraḷīyam* and began gathering together those who "think differently," there were other magazines, other gatherings, and other causes. And just as the biographical departures of environmental justice activists were predominantly departures from the CPM, so also the genealogical roots of people's protest activism, and of *Kēraḷīyam* magazine, could be traced to a decades-long effort to produce a viable leftism outside the institutions of Communist party politics.[11]

From the time the Communist Party of India emerged as a dominant player in Kerala politics, some writers and intellectuals had published small journals that, though often left-oriented, rejected the ideological uniformity promoted by Party leaders.[12] From the beginning, these "little magazines" offered an alternative space for intellectual exchange and fellowship.[13] Particularly influential in this respect were the magazines of M. Govindan, an intellectual disciple of M. N. Roy. Roy had been a founding member of the Communist Party of India, but had eventually divided with the Leninists over their emphasis on party discipline over intellectual freedom.[14] Like Roy, Govindan's rejection of communism was also a rejection of political parties as a mechanism for change.[15] As an alternative, he sought to use little magazines as a basis for discussion, seeking to generate social progress through the exchange of ideas.[16]

S. G. Mash, one of the few living activists who had participated in these discussions firsthand, described to me how Govindan used to travel throughout Kerala, stopping for a few days in each major city to hold discussions with young intellectuals and artists. While Govindan and his followers generally espoused some vision for socialism,[17] what distinguished these discussions was the diversity of views expressed. "In general, he just pushed people to think in new and different ways," S. G. Mash recalled. "He introduced a lot of young people to one another and got them imagining new things. All kinds of people. There would be discussions between people of every sort and persuasion; new-thinking people. And through that would come fellowship (*kūṭṭāyma*)."

Aside from S. G. Mash, few of Kerala's environmental justice activists were old enough to have participated in Govindan's magazines and discussion groups. But many of the older activists recalled participating in a similar social scene a generation later, in a rented room called Vanchi Lodge in downtown Thrissur. From the mid-1970s to 1992, Vanchi Lodge served as a workplace for leftist literary magazines and a social hub for "new-thinking" people.[18] Thrissur is centrally located along Kerala's major roads and railways, and Vanchi Lodge operated as a sort of rest stop for activists and intellectuals passing through. Jerry, an activist who famously made Vanchi Lodge his home for several years, recounted that the discussion there had been so perpetual and earnest that he had hardly ever slept. The magazines published out of Vanchi Lodge represented shifting ideological trends—from *Prēraṇa*, which was the semi-official literary magazine of Kerala's largest (but still marginal) Maoist Party,[19] to *Pāṭhabhēdam* , one of the first magazines to cover the budding environmental justice movement. And there were gaps during which no magazine was published. But according to Jerry, the discussion continued through it all, giving coherence to the role of Vanchi Lodge as a forum for an alternative leftism throughout this period.

Looking back, some older activists said that the founding of *Kēraḷīyam* magazine in 1998 had filled a hole left by the closing of Vanchi Lodge and its magazines a few years before. A few people even said that this was the very same community. Some of the same faces from Vanchi Lodge could be found in the *Kēraḷīyam* office. So could talk of solidarity with the people, which had origins in the Maoist vision of *Prēraṇa* but carried through to subsequent Vanchi Lodge magazines.[20] But more than either of these threads of continuity, activists pointed to similarity in the kind of community that *Kēraḷīyam* sustains.

The magazines of Govindan and Vanchi Lodge do not, however, represent the only genealogy of today's environmental justice activism. Contemporary activists also claimed historical roots in the conservation activism of the 1970s.[21] When the magazines published out of Vanchi Lodge were thick with Maoist rhetoric about "people's politics," John C. Jacob and his Society for Environmental Education in Kerala (SEEK) were trying to raise awareness about the deleterious effects of pesticides, pollution, and deforestation.[22] Jacob's organizing efforts also employed small magazines, some of which are still published today.[23] But activists remember him more for taking youth on overnight excursions into the forests to learn about ecology and connect with nature. Reciting poetry and singing songs about rivers and forests, budding environmentalists would bring to life (for a night) the environmentalist vision on the pages of their magazines. Decades later, Sunny's rain camp would give life to *Kēraḷīyam* in a similar way.[24]

The activists of Vanchi Lodge and SEEK differed vastly in their ideologies and aims, but they shared a practice of integrating magazines and small, face-to-face gatherings to develop an alternative (*badal*) mode of politics—a politics that avoided the social distinctions and organizations that dominate democracy in Kerala. In both cases, the magazines and the gatherings worked in tandem, with debates on the printed page feeding face-to-face discussion and vice versa. This looping discourse not only propounded various ideals, proposals, and strategies for change, but also stitched together a community of, in S. G. Mash's words, "new-thinking people."

In *Kēraḷīyam*'s office, one could see this looping process in real time. Located only a few blocks from the Thrissur railway station, the office

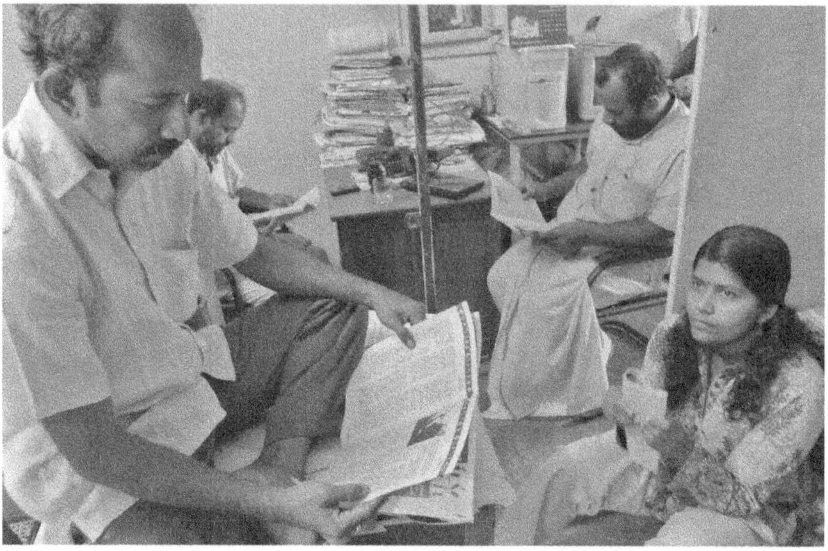

FIGURE 5. Front of the *Kēraḷīyam* office. Activists read and discuss the latest issue while folding copies to mail to subscribers.

FIGURE 6. Back of the office. Adarsh edits the next issue of *Kēraḷīyam*.

was sometimes called the station's "fifth platform" by its frequenters. As in Vanchi Lodge, a steady stream of passers-through fed ongoing discussion in the front of the office, the walls of which were lined from floor to ceiling with stacks of small-circulation magazines, both thriving and defunct. While some talked, others would pick up a magazine and page through. Unless he was out conducting interviews or taking a tea break, Adarsh could be found at a desktop computer at the narrow rear of the office, making phone calls, writing, designing the layout for the next issue, and trying not to hunch his sore back. He collected articles from those who came through the office or, when the grind of monthly publication permitted, sat with them to discuss ideas for future issues. When a new issue was ready, activists would crowd into the front of the office to fold copies and affix stamps, chatting as they prepared to send it out to *Kēraḷīyam*'s seven hundred subscribers across the state. Both the writing on the pages and the speech in the dim, dusty air sustained an ongoing conversation about what was wrong with the world and what should be done.

The looping circulation of discourse between print and speech—the back of the office and the front—can be understood to produce a shared vision for change, a common cause.[25] But there is no shared creed or official platform to be found at *Kēraḷīyam*. There is no membership list. There is a list of subscribers, but subscribing to the magazine does not mean one subscribes to all its views. Rather, as a publication and as a community, *Kēraḷīyam* is meant to be a platform for the kind of political activity that Dhanya believed impossible in the Communist Party or in any organization—a platform for "just continuously discussing and making modifications." Ideally, at least, it is meant to be a space of intellectual and moral freedom.[26]

## FINDING THE PEOPLE

The freedom activists sought was, in part, a personal matter. For Preethi, it was freedom to leave her stable career as a public school teacher. For Faiza and Adarsh, it was freedom to marry one another. For Dhanya, it included freedom from gender norms: freedom to take the reins in group discussions, to smoke, to go on extended tours with her documentary film collaborator, a married man. These freedoms were, on a day-to-day basis, what made life as an activist desirable and fulfilling. Yet, taken alone, they were not understood to be the real point. Everyday expressions of freedom were meaningful because, in activists' eyes, they contributed to a broader politics. And that politics was not centered on activists' lives. It was a politics of solidarity with the people (*janaṅṅaḷ*).

Activists' narratives of departure begin to answer this chapter's central question of how people make a life aimed at a larger cause. By setting aside purposes and roles tied to their specific communities—especially, the communities into which they are born—activists open up a space for more expansive aims. But taken alone, this is only a negative account of what activists leave behind; it does not explain how they arrive at new purposes and new lives—specifically, at lives aimed at solidarity with the people. Where do such larger causes come from?

There is more than one way to answer this question, and one's answer depends on the cause one pursues. Other possible answers can be found on the pages of the literary magazines of M. Govindan and the conservationist magazines of John C. Jacob, each attuned to their own expansive aims. Like the politics of solidarity with the people, Govindan's vision for freedom from the structures of parties and other organized groups was never imagined as only personal freedom. By opening minds to new thoughts, poetry and plays were meant to kindle a cultural renaissance that would, ultimately, transform politics as well.[27] John C. Jacob's magazines and camps arguably sought an even more fundamental transformation: a politics of unity that overflowed the boundaries between humans and nature.

For the environmental activists who practiced it, solidarity with the people went beyond both of these traditions. Govindan's magazine-based intellectual communities developed agendas for challenging class and caste hierarchies, but these were the visions and agendas of their own communities, dominated by highly educated elites. Likewise, while the conservationists' vision for harmony with nature was similar to that of solidarity activists, the circle of those who had developed these visions was very small and, arguably, highly exclusive. By contrast, through solidarity with people's protests, environmental justice activists sought to expand the circle of those who contributed to producing their vision for change. In doing so, they built on the Maoist people's politics of the Vanchi Lodge tradition.[28] They sought to extend their cause beyond the human-centered politics of the Maoists, and of Marxism more generally, by fighting for nature. But they also sought to go beyond prior environmentalisms by fighting alongside the people.

This notion of solidarity as a more inclusive and expansive environmentalist framework than conservation was crucial to the Dialogue

Journey, which was billed as a second run of another environmentalist trek along much the same route twenty-five years before. The 1987 "Save the Western Ghats March" (*paścimaghaṭṭa rakṣāyātra*) was remembered by many as an inaugural event in Kerala's environmental movement. This earlier march sought to educate the broader populace about the importance of preserving the rainforests by giving speeches, putting on plays, and performing poetry in mountain villages. From the perspective of the Dialogue Journey organizers, however, this awareness-raising effort was highly elitist. They sought to distinguish their own event from this tradition by explicitly foregrounding a two-way, egalitarian mode of communication with residents—a dialogue.[29] Rather than performing poetry and putting on plays, they stopped at homes and shops and conversed with local residents, emphasizing their desire to listen to residents' own perspectives and concerns. Their motto was "The ear is our tool, not the mouth." More importantly, by organizing the quarry coalition, Dhanya and others sought to facilitate the efforts of those mountain residents who were already resisting ecological destruction. Through solidarity with these protests, they reached for a mode of politics that not only was for all, but with all.[30]

Yet the politics of solidarity with the people posed a conundrum. Where could radical environmental activists find the people? In a different way from the literary politics of M. Govindan and the conservationism of John C. Jacob, radical environmental justice activists' search for an alternative politics depended on people other than themselves. This difference became apparent at the end of the Dialogue Journey. All of the local leaders were ready to form a coalition, but initially they wanted Dhanya to be its leader. After all, they said, she had initiated this process; she had gathered them on this hill. But Dhanya refused. She and other activists like her could only offer support, she insisted; the leadership had to come from the people themselves. If living for environmental justice was about fighting for the people, then solidarity activists needed the people to be out there fighting for environmental justice by fighting for themselves. They needed the people to not only fight, but to lead. But which people should count as "the people"? And how could following their lead also align with radical environmental activists' own vision for environmental justice?

### *Following the People*

In the everyday work of solidarity, this conundrum did not seem like a conundrum at all. Solidarity activists like Dhanya did not need to ask where they would find the people because the people were rising up all over Kerala, and they were crying out for support. As reflected in the Dialogue Journey, the practice of solidarity with people's protests consisted of a kind of political nomadism—of going where the protests were. Just as they visited quarry protests before and during the Dialogue Journey, so radical environmental justice activists also visited protest movements against pesticide use in cashew plantations, factory pollution along Kerala's many rivers, or road construction projects that displaced farmers or destroyed forests. They supported these protests in many different ways: helping distribute pamphlets and attract mass media, consulting on legal issues and giving advice on strategy, holding seminars on pollution, adding their bodies and their voices to rallies, marches, and roadblocks. Occasionally they would form an official solidarity committee, but more often their support was only loosely coordinated via text groups, email, and Facebook. In the absence of permanent, centralized coordination, the *Kēraḷīyam* office was a hub for the flow of itinerant solidarity activists, while the magazine's pages fed off this flow, collecting and collating news of the latest developments, reflecting activists' activity back to them in real time. And for activists closely connected with *Kēraḷīyam*—that is, from within this whorl of bodies and words—there was no question of where to find the people, only an urgency to follow them.

When I was following Adarsh, Dhanya, and other activists connected with *Kēraḷīyam*, this is how people's protests appeared to me as well. As I went from bus to bus, protest to protest, trying to keep up with the constant, intense activity of these activists and, at the same time, trying to keep up with writing field notes, I never had to think about how to find the people. I was busy videotaping their marches and rallies, interviewing their leaders and opponents, tracing their histories in magazines and newspapers, sitting for long afternoons in their roadside tents, and even defending myself against police accusations that I had incited them to protest. For me, as for the activists I was studying, "the people" were self-evident. They were farmers who were being poisoned by polluting factories. They were indigenous tribes whose way of life was being destroyed by deforestation. They were villagers whose homes were being swamped by city-dwellers' trash. They were marching in the streets. They were fasting until victory or death.

Yet, at moments in my research, this certainty was unsettled. This generally happened when I stepped outside the ambit of *Kēraḷīyam*. A politician with the Communist Party of India (Marxist) insisted that the campaign against the gelatin factory in Manamur was not a people's protest; only protests backed by the Party were true protests of the people. A journalist told me that the term "people's protest" was all wrong; these should be called "life protests" because these people were fighting for their lives. Another journalist—the editor of a small magazine sometimes seen as a competitor to *Kēraḷīyam*—objected to talk of "the people" as outdated Marxist rhetoric. Environmental protests, he said, belonged to the category of post-Marxist "new social movements." But the most poignant objection came from Mohanan, an independent publisher who had edited a Maoist small magazine in the late 1970s and early 1980s, the period when Vanchi Lodge was buzzing and "the people" emerged as a central focus of the search for an alternative leftist politics in Kerala. Mohanan had given decades of his life to magazines, committees, theater, and protests on behalf of the people. But when I visited his home to learn more about this history, he questioned whether activist talk of being "for the people" could ever be more than a smoke screen for activists' own agenda.

"For the people! What's that?" he asked, chuckling as if he had just delivered the punchline to some great joke. And he went on to criticize both his own former people's politics and that of *Kēraḷīyam*. "We write and talk in ways that make the people think that what we're saying is not for them. We alienate them. We think we are much more special. What we write sticks in that specialness. It won't become common."

Mohanan was not just quibbling over terminology; he objected to talk of solidarity with the people on principle. And while his criticism was rooted in his experience of Maoist people's politics, not environmental justice movements, it resonated with a practical problem faced by Dhanya and others: a gap between solidarity activists' vision for people's politics and the actual activism of those whom they called the people. This gap could be heard subtly in the discussion of who should lead the quarry coalition. It became much more explicit, however, in the gelatin factory campaign in Manamur, where the solidarity efforts led by Dhanya had been plagued by conflicts over gender roles, meeting processes, and alliances with party politicians. Eventually, nearly all of the activists connected with *Kēraḷīyam* had ceased going to Manamur. Some even said, in hindsight, that the gelatin factory protest had not really been a protest of the people after all.

This gap between the idea of the people and the politics of actual people is not unique to people's politics in Kerala. Writing about crowd politics in Bangladesh, Nusrat Sabina Chowdhury notes that "the paradox of peoplehood begins with the very act of representing 'the people,' which has always been a fiction. Its very existence requires a suspension of disbelief."[31] In any politics of the people, the people are missing.[32] Any group of actual people is only a part of the people, never the whole; that all-inclusive collective is always just over the horizon.[33] Thus, if activists' talk of fighting "for the people" excludes many people, this is not necessarily, as Mohanan suggests, a sign of bad faith. It is endemic to the category itself. No people's politics can ever simply follow the people; it must constitute the people as well.[34]

For activists like Dhanya, the gap between "the people" and any group of actual people meant that solidarity could never be merely about going out and supporting the ongoing protests of the people. Rather, part of the task of solidarity activists was to produce the people as a mass actor to whom activists could lend their support. Certainly, there were numerous environmental protests taking place in Kerala. But these protests had diverse actors and agendas. They could be interpreted and classified in many ways. Even if many protestors presented their own movements as people's protests, their agendas did not always align neatly with radical environmentalists' own vision for change. However dire the plight of those protesting, making diverse campaigns against factories, dams, quarries, and landfills recognizable as protests "of the people" required work. And this work was part of doing solidarity.

### *Formatting the People*

If the atmosphere around *Kēraḷīyam* magazine was especially thick with people's protests—so much so that questioning this category seemed almost nonsensical—this was perhaps because the words and images on the magazine's pages, the discussions in its office, and the many other activities initiated by its associates contributed to producing "the people" as a central actor in a radical environmentalist politics. The editors and associates of *Kēraḷīyam* also self-consciously worked to define, elaborate, and promote their vision for people's protests. The articles and images on the pages of the magazine contributed to unifying diverse campaigns under a single banner (the people) and weaving their diverse agendas into a sweeping vision for environmental justice. They formatted the wide-ranging protests of many distinct groups of people into the common protest of the people.

The gradual dissolution of the solidarity effort in Manamur brought this formatting work to the fore. Late on a Sunday morning, I joined Adarsh as he set out to interview Vijayan, the leader of the Manamur Action Council. The magazine had been covering the Manamur campaign since its inception, but Adarsh felt that an update was urgently needed for the next issue. By this time, the official Solidarity Committee had already become inactive. Some solidarity activists were still working with the Manamur Action Council, but Adarsh felt that energy in the movement was flagging, and he was worried by rumors of conflict between Vijayan and outside supporters, including Dhanya. Publishing Vijayan's interview, he hoped, could help sustain outside support and keep the local Action Council connected to the broader politics of people's protests.

Vijayan met us on the veranda, greeting us with his usual gruff, weary exuberance—his wide smile and his perpetually hoarse voice—and waving us onto a small couch in the airy, sunlit front hall of his home. He pulled up a plastic chair for himself while Ashokan, his close friend and a regular at the protest tent, brought each of us a cup of tea. Adarsh and I had tricky work balancing our teacups while also keeping our notebooks and pens ready for action—Adarsh taking down Vijayan's words and I taking notes on their interaction. Fortunately, we each had our recorders going as well.

The interview lasted over an hour, with Vijayan going into complex details of industrial engineering, legal process, and local politics. Adarsh asked him some pointed questions about disagreements with those involved in the solidarity effort. Previously, I had watched Vijayan evade such questions from both Adarsh and myself, but this time he candidly addressed criticisms that had been made of the campaign and his own leadership. Adarsh made many notes during these more controversial moments, but this was only the first transcript he would make. A few days later, I sat by him in the *Kēraḷīyam* office while he listened to the interview again, taking a second set of notes. The first notes were just a guideline, he told me, to aid him in making the second set, which covered about fifteen pages of a repurposed diary. The second set would be the basis for the update.

With each listening, each rewriting, Adarsh filtered Vijayan's words, selecting those that pertained to the purpose of the update. Sitting beside Adarsh at the computer, watching him type up and arrange the final columns of text, I could only catch glimpses of this filtration. Even so, it was not hard to see how the final article retained some particularities while reducing others, shaping Vijayan's words to fit a standard mold. The final article fit on a single page, printed on the back cover of that month's issue. It was structured by five bold-faced questions, with answers below. Each of the questions focused on recent developments: a court verdict allowing the company to repair its waste pipe, a recent meeting with the Minister of Industry, the discovery by the Action Council that factory waste was being sold to a company in a nearby village, and the campaign's stand with regard to the upcoming elections. The answers provided relevant details and commentary on each of these events.

While the specific information in the article was particular to the Manamur campaign, the sorts of questions and answers presented, the characters described (the Action Council members, the government officials, the company), and the hortatory style of language would all be familiar to *Kēraḷīyam* readers. Together, such elements conformed Vijayan's interview to a more general type of story about people's protest, a narrative format that made the Manamur campaign appear to be part of a common struggle and to partake in a common appeal for support from readers. On the pages of *Kēraḷīyam*, Vijayan and the hundred or so active participants in the Action Council become "the people" (*janaṅṅaḷ*), a collective that purports to include all.

## Thinning the People

What kind of universal collective is "the people"? As many scholars have noted, "the people" may seem to vaguely cover everyone, but "everyone" is always defined by contrast to one of several sets of other people (*āḷukaḷ*) who are not *the* people (*janaṅṅaḷ*).[35] In the protest update, one such contrasting set consisted of solidarity activists themselves—the update's intended readers—who were tacitly excluded from the people insofar as they were called to support them. More explicitly, Adarsh foregrounded contrasts with the protest's most direct opponents—in Manamur, the gelatin company and the police. Other key contrasts that establish the boundaries of the people include political parties, the courts, and government agencies. All of these have people (*āḷukaḷ*, plural of *āḷ*, "person"), but they do not have people acting as the whole of the people (*janaṅṅaḷ*).[36]

Making the Manamur Action Council the whole people, rather than simply one of many parts, depended in part on emptying the protest update of certain information. For example, beyond the pages of *Kēraḷīyam*, many actors claim to be the people. The gelatin factory workers and their families sometimes claimed the moniker for themselves while attempting to portray Vijayan and the Action Council as a front for typical "partial" actors such as political parties, wealthy foreigners, or even environmental activists.[37] I use "partial" in a double sense here; unlike the people, these other collectivities include only some part of the populace (partial, not whole) and they pursue an agenda favoring their own limited group (partial, not impartial). Adarsh's update did not, of course, mention opponents' claims that the Action Council was a front for other groups. But it also emptied the movement of other signs that might tie the Action Council participants to specific identity groups or organizations—that is, to culturally specific markers of partiality. Such editing work was most apparent in a "Protest Kerala" issue, a compendium of updates on major people's protests throughout Kerala. Here, there was no ambiguity about who the people were: they were listed in the table of contents.

This emptying work is also apparent in a map the magazine published of ongoing people's protests, on which each is designated by a dot. The names of the protests, likewise, are usually simply the names of places, only occasionally coupled with some descriptor of what is being protested ("Manamur protest," "Viḷappilśāla anti-waste protest"). The map strips campaigns bare of all particularity except their positions in space. In this image, there are no unions or NGOs, no castes or ethnic groups, no political parties, no environmental activists; there are only the people, rising up in locales across the state. One might say that, in this map, the magazine presents the people as a thin universal—capable of containing all protests because there are so few criteria for inclusion.[38] The protest movements appear similar because they have been reduced to such a skeletal form. They appear maximally inclusive because they lack markers of any of the usual exclusions.

## Thickening the People

Meanwhile, articles like the update on the Manamur campaign flesh out the details of campaigns to fit a common pattern of victimization, local mobilization, and a need for solidarity. This is most evident in the "Protest Kerala" compendium. The campaign goals represented in the compendium were extremely diverse—from efforts to shut down landfills and quarries, to demands for farmable land from the government, to attempts to ban alcohol from certain villages. But most updates had a common structure, progressing from a description of an injustice, to a story of the formation of an action council, to challenges such as opposition by police and government officials, and, finally, to a call to support the protest in its next phase. Such updates fatten out the bare dots on the map, or the names in the table of contents, giving every cam-

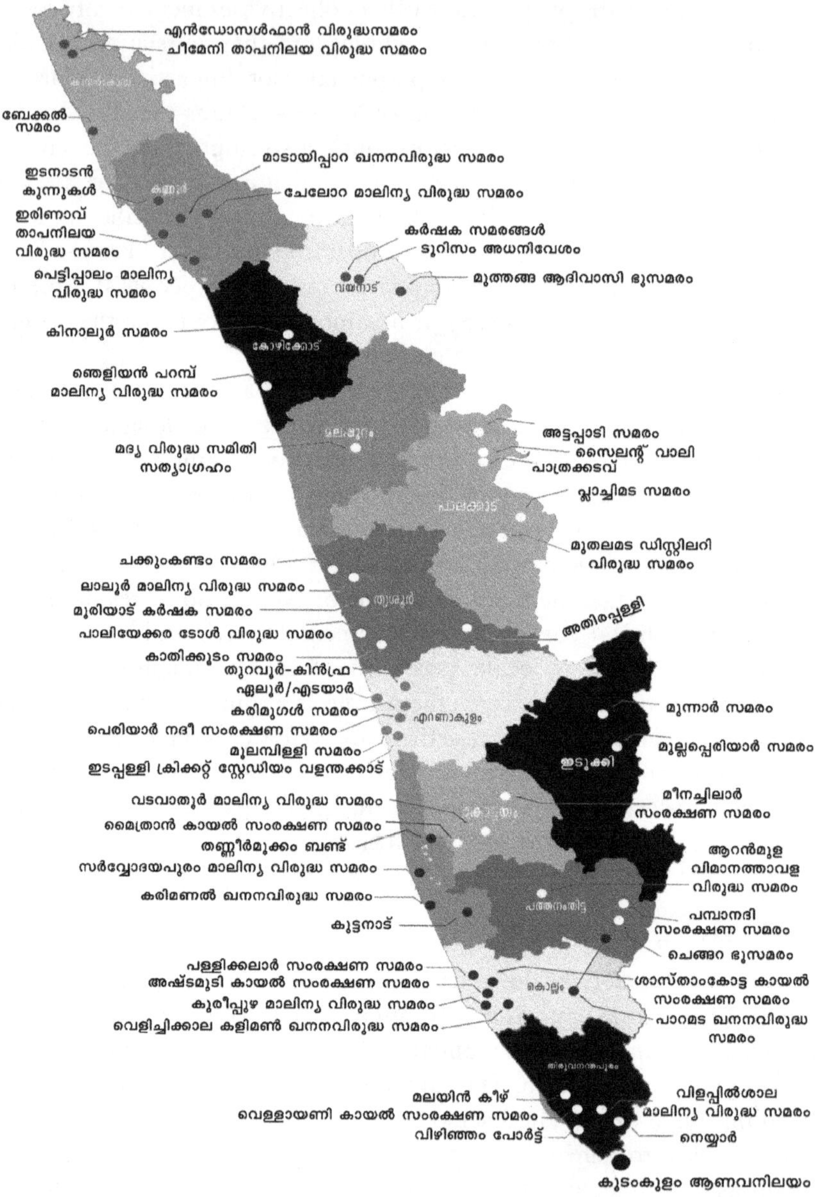


MAP 2. Map of Kerala's people's protests, published in *Kēraḷīyam*, August 2012. Courtesy of *Kēraḷīyam* and the Kerala chapter of the National Alliance of People's Movements.

paign the same flesh. Whether in Manamur or in dozens of other places across Kerala, the people described in these updates were living out the same story of struggle, and they were all in need of support.

This thickening work of the magazine extends the universalist agenda for people's protests beyond the concept of the people. The "Protest Kerala" compendium notwithstanding, protest updates are usually folded in among articles on methods of organic farming, conservation policy, biodiversity, recycling, alternative energy, and vegetarianism. There is also writing on topics that are less obviously environmental: criticism of mainstream education systems, reflections on the moral discourse of development, proposals for more radically democratic politics, and testimonials on the benefits of alternatives to Western medicine. The connection between these wide-ranging topics is not immediately clear to the untrained eye. And while the magazine sometimes includes an editorial, Sunny and Adarsh refrain from explicitly aligning the magazine with any overarching program, organization, or agenda. Though much of the writing has a manifesto-like tone, there is no one manifesto. Nonetheless, the magazine situates people's protests centrally within environmental politics, showing how the cause of the whole people contributes to the cause of even greater wholes. Seemingly distinct proposals for change are brought together in ecological interdependence, giving the sense—without any unifying creed or party platform—of a total vision for change that touches nearly every aspect of social life.

This integrative work is not solely a matter of agglomeration—that is, of collating and stapling articles on heterogenous topics into a single issue of the magazine. Rather, if all goes well with the editing, the right reader will be able to pick out the harmonies, the similarities of perspective across the diverse topics, and begin—as Sunny put it—to "envision things differently."

Yet this recognition of the pattern, this new vision, is not meant to be accomplished by the words on the page alone—not even by all the words on all the pages of all the issues of *Kēraḷīyam*. There are other magazines. There are documentaries like Dhanya's films on water rights and quarrying. There are WhatsApp groups, Twitter feeds, and Facebook posts. Just as Adarsh's protest update presented the Manamur Action Council as one instance of people's protest, and just as an issue of *Kēraḷīyam* situates people's protest as one battleground in a broader struggle for environmental justice, so also Adarsh and Sunny anticipated that readers of the magazine would situate its content within a broader array of alternative (*badal*) media. More importantly, they would situate it within their discussions in the front of the office, within their experience of meetings and marches in support of people's protests, and within the late-night singing and soaking of the rain camp. It was in conjunction with these activities that the vision for solidarity laid out on the pages of the magazine became a template for a form of community and a way of life. Formatting *Kēraḷīyam* and forming activists went hand in hand.

## FORMING THE FEW WHO FIGHT FOR ALL

*Kēraḷīyam* was not intended for just anyone. Sunny and Adarsh worked hard to expand readership, but that did not mean simply distributing it as widely as possible. It was not sold at local bookstalls, nor was it designed to attract ordinary people (*sādhāraṇakkār*). At the time of my research, another environmental magazine became popular for its stunning, often full-page photos of Kerala wildlife, and some *Kēraḷīyam* readers pushed Adarsh to adopt a similar aesthetic. But Adarsh and Sunny saw the aesthetics of *Kēraḷīyam*—plain newsprint in a black-and-white cover, with a few photos and abstract images throughout—as consistent with a decades-long tradition of small magazines that attracted by virtue of their ideas, not their looks. Adarsh and Sunny were not aiming to recruit as many readers as possible. They were cultivating what they called "serious readers" [Eng].

Despite their complaints about aesthetics, readers' accounts of how they read *Kēraḷīyam* fit with the editors' emphasis on substance over beauty, committed readership over mass appeal. Many subscribers—particularly those most active in supporting people's protests—treated the magazine as a kind of encyclopedia of contemporary people's protests and related concerns. They did not necessarily read each issue as it arrived, but they kept and collected them. Old copies of newspapers and mainstream magazines would be handed over to freelance recyclers who came by every few months, but *Kēraḷīyam* would pile up on a shelf or in a corner. Readers attributed this practice to the magazine's value as a documentary record of how campaigns had unfolded and what had been done—in many cases, to what they themselves had had a hand in doing.

### *Activist Inquiry*

While I most often saw such collections on the shelves of long-time activists, they were understood to be important to the process of becoming an activist, which was often described as *anwēṣaṇam*, or "inquiry." *Anwēṣaṇam* could involve a range of activities—including traveling to the sites of people's protests, attending seminars and environmentalist camps, and seeking mentorship from more experienced activists—but intensive reading was a central practice.[39] Those new to environmental activism borrowed old issues of *Kēraḷīyam* as well as other key alternative magazines and books. Both young and old strongly opposed the hierarchy and conformity of education in schools,[40] and the reading of *anwēṣaṇam* was largely understood as an independent endeavor. But this economy of lending meant that much of what came before the eyes of newer activists was selected by those more experienced.

Activists described such reading practices as leading to the new *bōdham,* or consciousness, that led to their involvement in people's protests and other efforts for change (see chapter 2). But *anwēṣaṇam* was not a transitional phase or rite of passage. Rather, *anwēṣaṇam* was ideally understood to continue even after one began to get involved in campaigns. For example, when I first met George he was intensively engaged in reading, including studying many issues of *Kēraḷīyam*, but said he was not yet ready to get involved as an activist. Two years later, I returned to find that he had helped to organize a protest against a factory near his neighborhood, yet when I suggested that now he had become an activist, he demurred, saying that really his focus was still on *anwēṣaṇam*.[41]

Similarly, for many of those I call activists here, continual *anwēṣaṇam* was their primary activity. They understood inquiry as the foundation for their commitment to the cause and their motives for their social action. Though others—such as the local protestors in Manamur—commonly referred to them as environmental activists (*pāristhitika pravarttakar*), many were reluctant to own this label. Dhanya, for example, protested that this term (specifically the "environmental" part) did not capture her "social and political activity," which she saw as less about promoting environmental values than about fighting the exploitative effects of capitalism on both humans and nature. More generally, many activists were as resistant to taking on ideological labels as to being members of any political party. As Adarsh told me on the day I first met him, they do not subscribe to any "isms." Their efforts for change were, ideally, ethically open-ended—a continuation of the processes of inquiry that brought them into the struggle for environmental justice.

*Serious Readers*

If Adarsh's role in *Kēraḷīyam* was to put out a monthly ink-and-paper template for an activist way of life, Sunny's role was to draw select people into that way of life and connect them with one another. Gatherings like Sunny's rain camp were crucial tools in this work, but most of these events were invitation-only. The work of finding and forming "serious readers" began in more marginal social spaces, where activist circles intersected with public life. The Dialogue Journey—tramping through the hills, discussing forest degradation and water scarcity with all who would answer the door—offered many such opportunities. And it was via the Dialogue Journey that Preethi, the woman who introduced herself as a former schoolteacher at the rain camp, was first introduced to Sunny.

On the final day of the Dialogue Journey, even as Dhanya was conversing with the quarry leaders on the smooth, grassy spot on the side of the hill, Sunny was down at the bottom of the hill, going after a different sort of quarry. He stood by a folding table just outside the double doors of a large meeting hall. The undulating, heavily punctuated melodies of impassioned speeches leaked through the doors, gushing up into intelligibility whenever someone passed in or out. Magazines had been spread shingle-like, the fore-edge of one tucked under the spine of the next, over the whole surface of the table. Dialogue Journey participants had carried these issues of *Kēraḷīyam* with them the whole way, offering them for sale but finding few buyers. The magazine had extensively covered the ecological issues affecting the lives of farmers, housewives, retirees, and other Western Ghats residents that activists met along the way. But these people were not the magazine's primary audience—nor were those protest leaders gathered with Dhanya on the side of the hill. The magazine was meant for the sort of people who walked in the Dialogue Journey, who had carried the copies to this table, and who were now, as the speeches dragged on, beginning to trickle out of the hall.

At first, Sunny did not seem to notice Preethi moving quietly down the table, scanning titles. He was mingling in the outflow from the hall, expertly performing a kind of conversational trapeze act—grasping the hand of one acquaintance, swinging away to the shoulder of the next, and introducing the two of them to each other as he twirled away to nab a third. But I knew what he was looking for, and I saw a chance to get a closer look at a sales pitch that I had observed, in various forms, many times before. I struck up a conversation with Preethi and made a swing of my own, guiding her into Sunny's path with a word of introduction. He shot a sly look at my field notebook, happy to play along. He asked whether he had not seen her somewhere before and, while she fumbled for a polite response (he hadn't), he drew the latest issue of *Kēraḷīyam* out of his bag. Had she heard of it? Yes, she said, but admitted she was not so sure she had read it. He pressed the magazine into her hands and began to praise its many virtues. It was all hyperbolic, all said with a laugh, but it was earnest nonetheless.

Preethi readily agreed to buy the magazine, but Sunny was after more. He wanted a subscription or, better yet, an article. Could she write a piece on her experiences on the Dialogue Journey? They could publish it in the next issue. "This is an activist magazine, not a mainstream publication," he told her. "We are trying to build a *kūṭṭāyma* (collective) of activists."

I had introduced Preethi to Sunny because I recognized in her the marks of a potential serious reader. She already had a passion for environmental issues—enough so that she had left her home village to join for portions of a fifty-day environmentalist walking tour through the mountains. Through participating in the Dialogue Journey, she had made friends with environmental justice activists and learned more about their causes and way of life. Yet she was still new to the scene, and I sensed that she was still looking to learn more. She had begun the process of becoming an activist. Sunny was eager to take her further along that path.

Yet, at first, Sunny's pitch did not seem to go very well. Preethi was quite willing to buy a copy of his magazine, but when he started to talk of subscriptions, authorship, and activist collectives, she only smiled quietly—and the more Sunny persisted, the more she shrank down behind her copy of the magazine, unresponsive. So Sunny tried a different tack.

"Have you met Father Sebastian?" he asked, reaching out and snagging the activist priest's hand as he went by.

She had, of course, since Father Sebastian was one of the leaders of the Dialogue Journey.

"What! You've met Father Sebastian and yet you don't know *Kēraḷīyam*? Sebastian is our top salesman!"

And though Fr. Sebastian hurried on without a word, his warm smile suggested that this was not entirely an exaggeration.

Sunny continued his pitch, "introducing" Preethi to one person after another: Ranjith, Manan, Amna and Amra ("What! You already know Amna and Amra? They are our brand ambassadors!"). All were people

Preethi had already met on the Dialogue Journey. Sunny wanted to recruit Preethi to be more than just a dues payer, and his attempts to draw ties between the magazine and Preethi's existing acquaintances were more than just another tactic to get her to reach for her purse. He was sketching out a web of relationships—a web in which she already had begun to have a place. One might say that he was building *Kēraḷīyam*'s "collective" (*kūṭṭāyma*) right there before her eyes, and he wanted to build her into it.

By the time Sunny pulled aside Saleem, whom she knew from her own hometown, she seemed convinced. I do not think Sunny ever got her to write an article for *Kēraḷīyam*, but a few weeks later, she told me that she had decided to leave her job as a schoolteacher—a stable, much-desired government job. She and Saleem, together with a few other activists, were planning to launch an environmentalist commune.

### *Activist Fellowship*

For radical environmentalists like Faiza, Dhanya, Adarsh, and Sunny, living for environmental justice meant putting values over relationships, community over cause. However, as was apparent in Sunny's opening statement at the rain camp, in practice this also meant fostering new kinds of community that were in line with activists' vision for social change. If Sunny's sales pitch was aimed at introducing Preethi into such a community, the rain camp she would attend two months later was meant to enact an ideal mode of community life. During the period of my research, activists associated with *Kēraḷīyam* participated in moonlight camps (*nilāv kūṭṭāyma*), rain camps (*maḷa kūṭṭāyma*), song camps (*pāṭṭ kūṭṭāyma*), and even gossip camps (*paradūṣaṇa kūṭṭāyma*), in which people would tell each other openly what they usually only said behind each other's backs. In these gatherings, activists put into practice many of the ideals proposed and discussed on the pages of the magazine.

Just as the content and ideals of *Kēraḷīyam* mix updates on people's protests with articles on diverse environmentalist and social concerns, so the values enacted in the rain camp exhibited a wide range of influences. John C. Jacob's tradition was clearly present—in the talk of organic farming and songs about harmony with nature, but most obviously in the very activity of sitting out in the rain. But the Vanchi Lodge tradition's search for a "people's politics" was evident as well, heard in Martin's news of the anti-quarry campaign from which he had just returned. And there were other threads: Surjith told of a conference he had just attended on the rights of sexual minorities; Sunny corrected a man for jumping in to introduce his wife, rather than letting her do it herself; late at night Xavier (Catholic by birth) cracked jokes about Christianity while Saleem (Muslim by birth) added in a few about Islam.

For some of those present, the camp was a taste of moral freedom—an opportunity to enact values that are censured or misunderstood in the broader society. I felt this too, lying alone on the beach after the introductions, watching the few wisps of cloud and wondering whether rain would, indeed, arrive that night. The clouds were unraveling and the waves had calmed, but the air was still keen with the rumor of a storm. Amna, one of Faiza's colleagues and a regular visitor at our house, came and lay down next to me, only a few inches away. I did not notice her just then. I did not notice her until the others came, telling us that they were headed further down the beach and inviting us to join. It was only then that I thought how strange it was to lie next to a woman on a beach in India. At night. As strange as sitting out in the rain with no umbrella. But Amna seemed to think nothing of it, and before long we were all sitting or lying beside each other on a plastic tarp—a tarp that could have been used to cover our heads—singing songs and waiting for the rain. Together, we were making present what was not present in the world around us, marking the changes that remained to be made and modeling what the world would be like when the work was done.[42] Here, women and men could lie on the beach together without fear of censure—without anyone thinking anything of it at all. Here humans reveled in the rain. Here, in a little way and for a little while, divisions between women and men, humans and nature, had been overcome.

The Malayalam term for environmentalist camps, *kūṭṭāyma*, connotes freedom as well. Activists use this same term for their own community—as in Sunny's "collective"—and, more importantly, to denote a kind of social bond. Unlike the *samudāyam* ("community") of caste and religious groups or the *kuṭumbam* ("family") of kin, *kūṭṭāyma* suggests a voluntary, non-hierarchical, and less rigidly bounded mode of togetherness. This is the term S. G. Mash uses to denote the "fellowship" forged through the early magazines and discussion groups of M. Govindan.[43] It is the term older activists used to describe the Thrissur-based network of activists connected by Vanchi Lodge and, later, *Kēraḷīyam*. It is also the term Adarsh and Sunny use to address their imagined readers: a group with no membership list, no creed, and no clear boundaries.[44] It is the ideal activist community: open to all and working on behalf of all.[45]

And yet, if camps are a place to enact freedom, that freedom also brings its own restrictions, pressures, and boundary lines. Sunny drew one such line when he corrected the husband for introducing his wife. That line was particularly palpable because, in his introduction, the husband had identified himself as an active member of the Communist Party. He was an old friend of Sunny's, but he and his wife left immediately after the introductions were concluded. And there were others who left quickly: a man who had made a documentary promoting the police and a woman who said very little in her introduction at all. Amna and Amra told me later that these people had left because they did not know anyone and were not really part of the group.

But those who stayed may also have found themselves on the wrong side of some lines. My awareness of lying beside Amna on the beach also brought back to me a conversation with Ahmed, my research assistant, about the Dialogue Journey. He had learned a lot and changed a lot, but he had always been uncomfortable at bedtime, when activists his age would lie close with no apparent regard for gender differences. He was not with me that night, but I wondered how he might have felt on the tarp with the others, getting soaked. Or how he might have heard Saleem's jokes about his faith. He and Saleem were both Muslim by birth, but here they would have been on different sides of the line. There were others on the tarp, I am sure, who felt more pressure than freedom that night. But I do not know who they were.

The tarp on the beach may seem far removed from the numerous tents and marches of people's protests. Far, even, from the hillside on which Dhanya gathered the quarry campaign leaders. The central actors in the politics of solidarity—the people—were not here. And there were some activists who thought such gatherings trivial or beside the point. One longtime activist in people's protests, a former fishworker who had gone on to lead a union of fishworker movements across Kerala, insisted that camps were a complete waste of time—endless discussion and no action.

Yet it was not hard to see how there, in the rain camp, the idea of environmental justice was more alive than anywhere else. If environmental justice was, in historical terms, the culmination of a decades-long struggle to find an alternative to the politics of group affiliation, then that alternative was most fully realized in camps. For one night, in a tarp-sized pocket of Kerala, living for environmental justice was not the alternative anymore. It was the norm.

## ARE WE NOT ONE?

Activists like Dhanya and Sunny positioned themselves as "different" in a rigidly communal society, a society of families grouped by caste, religion, and party. Yet their society was also awash in traditions that, at least in some guise, claim to transcend community boundaries. In breaking with the identities and communities into which they were born, environmental activists left behind not only Communism, but also Christianity, Islam, and Hinduism—each of which also claims, at times, to be open to all and to work on behalf of all. Environmental justice activists' commitments to "the people" and "nature" built on this cultural background; in this sense, these activists were not uncommon at all. In their view, the problem was that each of these projects, while claiming to include all and fight for all, had only become a new basis of social differentiation and group loyalties. These activists wanted the politics of environmental justice to be more than just another universalist religion or party. The real stakes in becoming "different" were not in a particular set of values, but in sustaining an approach to ethics that put values over relationships, cause over community. They were trying to reorganize their social lives in morally inclusive and expansive ways.

Yet radical environmentalists' own efforts to build community, as much as they emphasized freedom and inclusion, always tended to produce new forms of exclusion as well. This was apparent in the rain camp, where the rejection of Communist party politics and organized religion inevitably drew lines between those who belonged and those who did not. This problem became even more apparent to me one afternoon in *Kēraḷīyam*'s office, when a CPM member arrived to discuss a marriage proposal.

The proposal had come from Rohit, a close friend of Adarsh and a regular in the office. Rohit was only marginally an activist—I never saw him at protests or planning meetings—but his social life revolved around *Kēraḷīyam* nonetheless. He ate, slept, smoked, and mostly just sat around in the *Kēraḷīyam* office, and when it came time to find a spouse, he attempted to do that through *Kēraḷīyam* as well. True to form, Sunny took care of the social side, using his vast network to establish contact with the family of a woman whose online profile fit Rohit's criteria: smart, good-looking, and, above all, from a caste other than his own. Adarsh then helped Rohit compose a carefully worded profile, weaving together a suitable mix of subtly alternative, yet not objectionable, qualities and interests. The girl's father agreed to explore Rohit's proposal, and a few days later they were serving him tea at Sunny's desk.

But the meeting did not go as hoped. The father described himself as a person who "believes only in reason," and said this was why he had sought an intercaste marriage for his daughter. This had attracted Rohit to her profile; it had stated that she was not only open to all castes but, like himself, sought to marry across caste. But the more the father talked (and he did most of the talking), the more it became clear that Rohit would not be an acceptable candidate. The father would only approve of a thoroughly atheist family for his daughter, and Rohit's mother still practiced some Hindu rituals at home. The father said he was open to other ways of thinking, and the decision was ultimately his daughter's, but it was clear that religion was a disqualifier.

By the time he left, Rohit was discouraged, while Sunny and Adarsh were utterly fed up. As Adarsh and I rode the bus home together, he explained his frustration. "He is an idealist," Adarsh told me. "He wants everyone to see things the way that he does. It's the same with religious people: they have their ideas and they want to convince everyone else about them, even by force. But being an idealist shouldn't just be about trying to get others to see your own ideal. It should be about freedom."

What bothered Sunny and Adarsh was not the father's values—they shared his opposition to religion and, especially, to religious and caste endogamy. What bothered them was his approach to relationships. By insisting that his daughter only marry into an absolutely atheist family, he was apparently using his position as her father to force his values upon her. More broadly, he seemed to be using marriage to create a community of pure-bred atheists. This resonated with a joke I had heard in the *Kēraḷīyam* office about Communist Party members: they were once anti-caste, but they had become a new caste. The crux of the joke was the "Communist family," that only seeks to marry within the party. And like many jokes, it was funny because it was so familiar. In part because of Communist activism, many casteist practices have declined in Kerala, but marriage across caste remains rare. That is why Faiza and Adarsh's intercaste marriage was so important to them, and it is why Rohit sought such a marriage for himself. But the same principle of endogamy could make those who reject casteism into a caste of their own.

But were the radical environmentalists so different from the Communists they ridiculed? The radical environmentalists' criticism of the

Communists can arguably be seen as an extension of the Communist movement's own ideals. They posed "people's protests" as an alternative to party politics, but the rhetoric of "for the people" had its deepest roots in the Communist movement of the early twentieth century. Environmental justice activists laughed at the Communists for the same reasons that Mohanan laughed at the environmental justice activists: they found their rhetoric hypocritical. But they did not laugh at their cause.

The resonances with the Communists do not end there. The radical environmentalists' expansive moral project also produced a paradox that resembled the contradictions they saw in Communist politics: the broader the scope of their moral vision, the more uncommon their cause became. Rohit's difficulties finding a spouse illustrate this problem. From one perspective, his marital philosophy was maximally inclusive—he sought to transgress the restrictions of community boundaries by explicitly seeking an intercaste, interreligious marriage. Yet this approach to marriage also severely limited his prospects. He could only marry someone who was from another community and wanted to marry outside that community. This was a very small group. By seeking to include everyone, he excluded almost everyone.

Radical environmentalists in Kerala saw the politics of solidarity as one way to overcome this paradox, but it reintroduced it in another form. Not wishing to follow prior universalist movements in promoting a vision as elite as it is broad, they sought to ally themselves with a community that has no bounds: the people. Yet the very act of committing themselves to working for the people made activists, by definition, not the people. In pursuit of being one with all, they set themselves apart.

To expand their moral purposes beyond all community boundaries, radical environmental activists in Kerala cultivated a mode of sociality that put cause over community, values over relationships. As institutions, magazines facilitated this project by constantly foregrounding discussion and inquiry, rather than fixed tenets or membership lists. Magazines kept community boundaries loose. Nevertheless, as Sunny suggested at the beginning of the rain camp, this way of overcoming exclusions was all about being different—because nothing made activists more different than their efforts to live in solidarity with all. By the time the rain camp participants moved from the beach house to the plastic tarp, the few who did not fit in had already gone home. And when, in the wee hours, the rain at last began to fall, "the people" who gave the event purpose were far away, sleeping under their own roofs.

2

# Living for Our People

## FOR THE *NĀṬ*

"They aren't organizing for a 'cause' [Eng]," Dhanya told me. "For example, if they offered to take the factory's waste somewhere else, the people in Manamur would agree to that very quickly . . . but this waste will make the same problems there that it is making in Manamur. These people don't have that consciousness yet. I see that as this protest's biggest weakness."

By the time Dhanya told me this, she had already stopped going to Manamur, and so had most activists like her. Their reasons varied; many still seemed to be thinking through what had gone wrong in their collaboration with local protestors of the Manamur gelatin factory. As leader of the official Solidarity Committee (*aikyadārḍhya samiti*), Dhanya had been at the center of it all. She said that it was a matter of motives, of purposes. She defined "cause" as "the motor inside me, the aim." Without a cause, she said, the Manamur protest was weak and unlikely to succeed.

The Solidarity Committee had lasted less than six months. The previous May, hundreds of thousands of glistening fish bodies floated to the surface of the Neelajalam River, just downstream of the factory's effluent pipe. The Manamur Action Council had put out a call for greater support, and activists began arriving from across the state. In July, police charged at a crowd of protestors that had gathered at the factory gates—a mix of locals, solidarity activists, and curious onlookers—beating them with their batons and ransacking many nearby homes. The footage of collapsing bodies, bloody welts, and broken furniture had made the anti-gelatin factory campaign the most famous people's protest in Kerala. Solidarity activists had flooded in to show support, as had party politicians, religious leaders, medical professionals, and researchers. Hopes had been high.

But only a month later, when Dhanya gathered more than two hundred activists on a school ground near the factory for a Solidarity Convention, the mood began to sour. Very few members of the local Action Council, a dozen perhaps, attended this event. Some told me that, because of the name, they had thought the event was only meant for outsiders. Others noted that there had been a big wedding in Manamur that day. But Dhanya and other solidarity activists saw the lack of local attendance as a symptom of more chronic tensions in their collaboration with the Action Council. Over the next several months, I watched as the collaboration gradually came undone. Dhanya continued visiting longer than most, saying that she hoped to "raise consciousness" (*bōdhavalkarikkuka*). Adarsh published an update on the gelatin factory protest for *Kēraḷīyam*, described in the previous chapter, in the hope that this would help sustain interest in solidarity work in Manamur. But nothing seemed to work.

Dhanya's analysis of what hindered collaboration in Manamur may seem puzzling. How could it make sense to claim that those who lived so close to the factory—who had watched their children grow sick from the stench and the slime; who had marched, fasted, and received beatings by police in order to stop the pollution; and who had called out to activists like Dhanya for support—lacked a cause? Were they not living their cause every day? Yet Dhanya suggested that this was not enough; that the mere impacts of pollution were not a sufficient motive to energize and sustain a movement. She suggested that locals needed to fight for something more than an end to pollution in Manamur; they needed to fight pollution in principle. And her complaints resonated with how members of the Action Council described their own activism.

In the opening chapter of the book, I introduced Sunitha, who described her work with the Action Council as part of being a good mother and neighbor. Sunitha and others in the Action Council said their aim was to protect the *nāṭ*, a spatial term that, like the English "home," refers to the place where a person belongs. The referent of *nāṭ* is person-centric, varying depending whose *nāṭ* is at issue; when I fight for my *nāṭ*, I fight for a different place than when you fight for yours.[1] The Action

Council members were fighting to protect their families, their neighborhoods, their village, their home. And the limits of this commitment came through clearly several months after Dhanya stopped going to Manamur, when it seemed the hypothetical scenario she had posed might become a reality. There were rumors that the multinational corporation that owned the factory was considering relocating it to the state of Gujarat. I asked Vijayan—the leader of the Manamur Action Council—what local protestors planned to do.

"Our stand is that here, in the circumstances of our village, this factory cannot be allowed," said Vijayan. "Where they'll take it or what they'll do with it tomorrow is not our issue."

What does it mean to commit oneself to fighting for one's own community? When compared with radical environmentalists' expansive vision for environmental justice, Vijayan's stance on moving the factory elsewhere may seem narrow or self-interested. It may seem like what some call NIMBYism or "Not-In-My-Backyard" activism—a mode of politics that puts self-concern over concern for others and is often seen as fundamentally at odds with environmentalism. Alternatively, it may seem like what we expect of those directly impacted by industrial pollution—victims of injustice, for whom fighting for one's self-interest is justified and even praiseworthy. In either interpretation, critical or approving, Dhanya's complaint rings true; fighting for the *nāṭ* is not a full-fledged activist cause—not a larger moral purpose like the vision of environmental justice pursued by the solidarity activists. In this definition of "cause," fighting for one's own community can only be a cause when one's ultimate aim is to save the world.

This chapter refutes both of these readings of the motives of people like Sunitha and Vijayan. Whereas the radical activists sought to disconnect their cause from community ties, to *live for* without *living from,* the local activists sought to combine *living for* and *living from* synergistically, to make their local community their cause. More than a geographic designation, protesting as one of the "locals" (*nāṭṭukār* or "people of the *nāṭ*") meant calibrating one's purposes to degrees of relationship, prioritizing some aims while deprioritizing or excluding others. Specifically, local protestors excluded many of the "broader" aims of Dhanya and other solidarity activists. And this, ultimately, led to conflict between the two groups.

But this did not mean that locals in Manamur were NIMBY activists, motivated only by self-interest, nor that their activism was merely a reaction to the suffering caused by pollution. Instead, a deeper understanding of their activism forces us to question the dichotomies between NIMBYism and environmental activism, interests and ethics. While the local activists' approach to activism was grounded in their experiences of living with pollution—in coughing fits, rashes, and cancers—it was no less value-based than the lives of solidarity activists like Dhanya or Faiza. Moreover, just as the solidarity activists questioned the seeming narrowness of the locals, so the locals activists' person-centric activism offers a critical perspective on the social detachment presupposed by expansive moral projects like that of the solidarity activists. By keeping both perspectives in mind, we can better assess the possibilities and limitations presented by various ways of living for environmental justice—or, indeed, for any activist cause.

## LIVING WITH THE SMELL

When members of the Manamur Action Council told me how they joined the protest, they talked about "the smell."[2] The smell was both an actual smell—dank and full-nosed, like rotting flesh, but with an acrid, chemical edge to it—and a word to capture the whole experience of pollution in the air, water, and soil. It was the soapy film on the surface of the wells, the itchiness after a dip in the river, the wilted banana trees, the coughing fits at night. Like all pollution, the smell was as much a social condition as a bodily experience.[3] Loss of property value was the most concrete, quantifiable manifestation of this, but the smell seeped deep into social life. Mental health problems and alcohol addiction had increased, I was told, because of constantly breathing the smell. Stories of neighbors or relatives driven away by the smell were common. Again and again, Action Council members recounted the tale of a wedding reception that had to be canceled because the groom's family had refused to come to such a smelly place.

Such stories of the smell can make local protestors' motives seem self-evident—no one would want to live like this. There may be no need for a story about ethical values; their activism might be an organic, automatic outcome of the conditions of their environment. The same stories can be retold as stories of self-interest or NIMBYism. After all, self-interest—especially, economic interest—was front and center in the tales of past protests of the gelatin factory told by protestors and other Manamur locals alike.

By all accounts, for as long as there has been a gelatin factory in Manamur, there have been protests. The factory was a joint venture between a multinational corporation and Kerala's State Industrial Development Corporation, and some still told me its arrival brought opportunity and progress.[4] But older villagers recalled that, even before the factory opened in 1979, an affluent Nayar businessman had warned that it would destroy the land.[5] No one had listened to him, they said, because everyone but this man had wanted a factory job. Some did get jobs, but these came at the cost of a miasma that billowed out over the factory walls and a slime that crept into the wells.[6] Over the ensuing thirty years, various villagers organized sit-ins, speeches, and petition drives in protest, but all of these eventually petered out. People said the factory had paid them off. Participants in the Manamur Action Council said their campaign was different; they and their leader Vijayan would never betray the *nāṭ*. But for these other protests, at least, the cause had only been a matter of how much the company would pay. And many in Manamur figured it was only a matter of finding the right price for Vijayan and the Action Council as well.

But the smell was only part of the story of how people became protestors. Many of those who dealt with the smell were not involved in the protest, and only a small minority of Manamur residents were as active as Sunitha. Even Sunitha had lived with the smell for some fifteen years before she became an activist. During those years, she said, she had been an ordinary housewife, raising two young girls on the ground story of her house while the workers from the family jewelry-making business pounded gold on the story above. The smell was an irritant, but not an unendurable one. When her uncle retired from his job as a physician in the city, she convinced him to move to Manamur, pitching it as an idyllic village where he could stave off boredom by volunteering at a nearby hospital. This same uncle would go on to warn her about the dangers of the smell, tying it to high rates of cancer in the area. And when Vijayan and others started the Action Council in 2008, her husband soon got involved. But Sunitha stayed home. She was a busy mother.

"Up until then," Sunitha recalled, "I was just a woman who stayed home as a housewife, looking after her children and husband. I had [done] nothing like this. . . . Our typical woman who just looks after her family—that was me."

What made Sunitha become an activist? She told me it was a seminar that had been organized at the local Manamur library exclusively for women. The organizers were several environmentalist women from outside the village—solidarity activists like Dhanya. Up until this time,

FIGURE 7. Protestors in Manamur gather at the factory gates as a politician makes a speech in solidarity.

most local protestors in Manamur had been men; the solidarity activists wanted more women like Sunitha to join the local Action Council as well. And Sunitha said their intervention persuaded her.

"After getting this awareness, the mothers here began to come to the forefront of the protest. After that, we had awareness about the bad effects on our village (*nāṭ*), and we mothers joined [the Action Council] in order to save our village (*nāṭ*)."

The library seminar sounds much like Dhanya's own consciousness-raising efforts in Manamur, which had also focused on increasing women's involvement. One might say that through the library seminar, Sunitha gained both consciousness and cause. Moreover, Sunitha's account also resonates with pedagogies of activist "consciousness raising" around the globe, in which the oppressed come to see their experiences of oppression in a new light.[7] But Sunitha's account also differs from Dhanya's notion in crucial ways. First, the new "consciousness" mothers gained was an awareness of the bad effects on the *nāṭ,* and their newfound purpose was to save their *nāṭ.* Second, in the context of Sunitha's broader story of living with the smell, the awareness she gained at the library seems to be less of an awakening to injustice, or even to knowledge about the factory's pollution, and more of a new determination to do something.

This second difference is captured by a subtle contrast in each woman's terminology: the root word Sunitha used to talk about her experience of gaining "awareness," *bōdhyam,* is similar both in sound and meaning to Dhanya's term for the lack of "consciousness" in Manamur, *bōdham.* In certain usages, the two can overlap.[8] This was the case in solidarity activists' talk of "raising consciousness," and it is possible Sunitha also meant it in this way. But Sunitha's phrasing may also express a new moral conviction to take action, to act rightly, without necessarily suggesting a moral awakening. And this seems more consistent with how Sunitha and other Manamur protestors described their entry into activism as well as their efforts to recruit others. Sunitha had known about the pollution for a long time; her physician uncle had talked with her about its health effects, and she had no doubt spoken with her husband about the Action Council's work. But she had been busy being a mother. Now—because of the seminar but also because she was a mother—she felt she had to do something.

This does not mean that new knowledge, including that offered by solidarity activists like Dhanya, was not important to Manamur protestors' convictions. In Sunitha's case, the library seminar clearly helped her to see her experiences in a new light. Likewise, other Action Council members talked about the value of the knowledge offered by solidarity activists—including scientific knowledge as well as knowledge of how to conduct people's protests. For example, another prominent participant in the campaign, Mary, saw solidarity activists' knowledge as important not only to protestors' own convictions but also to their efforts to recruit others.

"The environmental activists are able to understand much more scientifically the problems the company is causing in our area. People who are ordinary people don't know, scientifically, how and why [these problems] are happening. We can learn a lot when people like this come, and then, even when they aren't working here, we can tell the people what they said and persuade them (*bōdhyappeṭuttuka*) as well."

As much as they valued solidarity activists' knowledge, however, members of the Manamur Action Council, including Sunitha and Mary, grounded their decision to fight for the *nāṭ* in the continual, everyday misery of living with the smell. And visiting Manamur day after day, I saw many moments when *bōdhyam* seemed to be happening—moments in which activists reminded one another of the effects of pollution and stoked their conviction to act. In the long, hot hours of perpetually occupying the protest tent, the smell (the literal scent on the air) was a stimulus to new conversation. If they caught a whiff of it wafting over the factory wall, they would wrinkle their noses and note that the factory must be operating. If it seemed strong, they would say the factory was increasing production. But they would also talk about how, on average, the smell was weaker now than it had been in former days—and this was a sign that they were winning the struggle. Likewise, they commented on the trucks that entered and exited the factory gate—another gauge of the relative strength of production and protest. Sometimes they would stand in the road to block the trucks, attempting to turn them back or, at least, to sniff out the telltale scent in the gray liquid leaking from the back. Once, they attempted to engineer a partial barrier in the river so as to direct the factory's effluent back into its own pumphouse, believing it could not operate with the polluted water. This project meant wading waist-deep in the mucky river for hours on end, and many complained afterward of itching and rashes.

In all of these practices, as in Sunitha's story, we can begin to see that the smell was not simply something that happened to Action Council members; their activism also shaped and amplified the smell's effects on their lives. Not that they chose to live with the smell in the sense that the radical environmental activists chose to reject religion, marry across caste, or start a commune. But fighting for the *nāṭ* meant attuning to the smell and, in some cases, amplifying it.[9]

This process could be a virtuous cycle—attunement to the smell could lead to stronger conviction, which could lead to blocking trucks

FIGURE 8. Protestors sniff and scrutinize a leaking truck that has just emerged from the gelatin factory.

or damming rivers, which could lead to further attunement, and so on. But it could also bring about intense despair.[10] One day, I found Vinod—a large, energetic man who loved to be at the head of marches and the center of parties—sitting alone and despondent at the edge of a paddy field. He and his wife had thus far been unable to have children, and while their doctor was uncertain about the underlying problem, Vinod was not; he was convinced that the factory's pollution had given him cancer. Like Sunitha, he had heard medical experts associate the smell with high cancer rates in the area, and he had seen it gradually destroy the bodies of his friends and neighbors. He said he knew he should just move away from Manamur, but he could not bring himself to go. Thus far, he had avoided mentioning it to anyone, including his wife, and he asked me not to tell anyone either—not until he was able to get tested. Ultimately, the tests came back negative, and this burden eased. But the memory of that day stuck with me: Vinod, believing the smell was killing him, had still been determined to stay in Manamur and fight.

Bodily attunement was not the only way that becoming an activist multiplied the smell's effects. Some activists, especially those who worked as day laborers, lost wages for the time they spent marching or occupying the protest tent. Housework went undone and children fell behind on their schoolwork. Many were accused in legal cases, both criminal and civil, for their roles in protest activities; there were court dates, legal fees, and fines to contend with, not to mention the shame of being put on trial, however unjustly. There was also, for the most committed, the very strain of the constant, frenetic activity required to sustain the fight. Early in my visits to Manamur, Vijayan seemed to have unlimited energy and zeal, constantly zipping on his motorbike from one place to another and sleeping only a few hours a night. But three years later, he had developed a range of vague ailments—body aches, kidney trouble, hypertension and high sugar—that sometimes kept him in bed for days. "I can't rest," he said. "During crucial periods of the protest, that's one problem I have. This is a time like that—a most crucial time—so I have that tiredness, that awful tiredness."

Even as the smell drove some residents of Manamur to join the Action Council, choosing to join also led them to a more intense experience of the smell. The impacts of the smell increased not only as one came nearer to the factory walls, but also as one moved closer to the center of the fight to shut the factory down. The more time people spent in the tent, the more aware they would be of fluctuations in the flow of stench over the walls and the flow of trucks through the gates. The more people marched, blocked trucks, or battled with police or laborers, the more they were likely to accumulate court fees, fines, and even jail time. And the more the demands of the fight dominated people's daily activities, as they did for Vijayan and a few others, the more the stress seemed to wear them down. This was what it meant to become an impacted community.[11]

## CONFIGURING COMMUNITY

Life as an activist mother was different from Sunitha's earlier life, her "housewife" life. Once she joined the Action Council, Sunitha began to go to the protest tent beside the gelatin factory gates. The tent was the center of the protest's activities. It was where protestors began marches through the roads, lay on cots during relay fasts, and discussed their next moves; it was where visiting politicians and celebrities gave speeches and were photographed accepting flowers; it was where Sanoop gave a feast of beef and ghee rice to celebrate Eid-al-Adha. For Sunitha, however, the protest tent was most of all a place to sit and chat with other protestors in the evenings, after dinner, often well into the night. Here, Sunitha found a community of people like herself, for whom activism was an extension of their existing roles in their communities. But being here also put her in new roles—not only her, but all of the women who came out to the tent after dark.

"Have you ever seen anything, anywhere in Kerala, like this before?" she asked me, "Women meeting up together in the middle of the night? Have you, John?"

"It's rare," I said, thinking of women like Faiza and Dhanya, for whom it was not so rare at all.

"It's certainly not done!"

Sunitha's fight against the factory was rooted in belonging to a place. But it also led her to occupy that place in new ways, giving new meanings to being a mother and a neighbor. For Action Council members, seeking synergy between *living for* and *living from* was primarily a matter of grounding the former in the latter—making their community their cause. But just as protesting the smell reshaped their sensory experience of living near the factory, so also fighting for their community reshaped that community, changing what it meant to belong to the *nāṭ*. Joining the Action Council made Sunitha local in ways she had not been before.[12]

And yet, Sunitha's activism did not make her local in every way. Fighting the factory strengthened Sunitha's ties to some neighbors, especially the other women on the Action Council, but it also alienated her from many others. Sitting in the tent outside the factory gate—located on Manamur's main road, right in the heart of the village—was an unavoidably public act. Some of Sunitha's neighbors supported her, but many—especially those who had family members working in the factory—argued with her, telling her to stop protesting. Conflicts with her neighbors increased dramatically when, not long after joining the campaign, Sunitha decided to run for a seat on the *pañcāyatt*—a local governing body like a town council. Several *pañcāyatt* seats were reserved for women, and the anti-factory Action Council hoped that, if their members got seats, they could use the *pañcāyatt* authority to force the factory to stop polluting or shut down. Running for a seat thrust Sunitha even more into the public eye—now she was one of the main voices for the campaign. This meant more arguments with her neighbors and even with some of her family.

"They said so many things to try to get me to withdraw," she recalled, "But I kept on campaigning, without abandoning the protest. When they asked what party I was with, I told them, 'We are running as independent candidates for the *nāṭ* and for the good of all, to stop the company's evil acts and protect the river and everything.' And they immediately told me, "No votes for you!'"

## *Managing Multiple Affiliations*

Sunitha's framing of her retort to her neighbors is revealing. It shows that being a local activist meant carefully managing one's affiliation with other groups, such as political parties. Sunitha pointed out to her critics that she was not running with any political party. This did not mean that she did not belong to a party—in fact, she was an ardent supporter of the BJP, the Hindu nationalist party that had often been criticized for divisive, anti-Muslim politics. She was open about this. But in representing the fight against the factory, she took a different footing, positioning herself as someone who was fighting for all local people—not just Hindus, not just a particular party. She did not deny her membership in a party, but she bracketed it, or set it aside, in order to fight for the *nāṭ*. Unlike the alternative politics of the radical environmental activists, here, working "for the good of all" did not mean fundamentally opposing the widely prevalent politics of group affiliations like party, caste, or religion. Rather, it meant accentuating one kind of group belonging—commitment to the *nāṭ*—by carefully managing other kinds.

Just as Sunitha bracketed her party affiliation, Action Council members often bracketed religious and caste affiliations as well. Consistent with the demographics of the area, most Action Council members were Hindu or Christian.[13] But when Sanoop, a Muslim, announced he was treating everyone to beef and ghee rice on Eid-al-Adha, it was talked about for weeks prior, and nearly everyone came, including high-caste Hindus. Sanoop did not abandon his religious affiliation, nor did his guests, but they celebrated together as they protested together. Likewise, people from a wide range of castes participated in the Action Council—including both Rani, whose ancestors had once owned much of the land in the area, and Sujit, whose ancestors had once been slaves on Rani's ancestral land. Rani still hired Sujit for agricultural labor sometimes. Both were active in the Action Council, but they participated in different ways. Sujit painted signs and banners for the protests, an activity associated with the traditional occupation of his caste as artists, while Rani mobilized assistance from her foreign contacts and occasionally gave public speeches. Protesting together bridged caste distinctions, but it also built on them.

FIGURE 9. Protestors chat in the tent with a Congress politician (center, in white).

Fighting for the whole *nāṭ* could sometimes mean highlighting, rather than bracketing, ties to nonlocal groups. The main leader of the Action Council, Vijayan, had been an active member of the Congress Party since his college days, and the Action Council's opponents often claimed that the protest of the factory was simply Vijayan's attempt to further his political career. Not only did Vijayan not deny his Congress affiliation, he actively sought the involvement of high-level Congress politicians in the protest—only he insisted this was all done to benefit the *nāṭ,* not his party. Likewise, other Action Council members argued that Vijayan's Congress activity was a strength, both an indication of his leadership skills and a source of connections to powerful people.

Similarly, Action Council members at times emphasized caste or religious identities as assets. Rani and her husband were treated with a deference that, while not explicitly casteist, was far from caste-blind. Rani, especially, was ascribed a kind of minor celebrity status by some, who referred to her as *tampurāṭṭi,* a feminine aristocratic title—not to her face, but not unseriously either. Sunitha once organized a bus pilgrimage to Hindu temples for Action Council participants. Only Hindus participated (Ahmed and myself aside), and on the bus we heard some complaints about other religious groups that would not, I think, have been voiced in the protest tent. Sunitha and other participants framed the tour as an activity that strengthened the protest through camaraderie, spiritual rejuvenation, and the hope of divine intervention. They appealed to the gods for healing from illness and high marks on school exams, but they also asked for help in saving the *nāṭ*. Just as Vijayan's Congress connections or Rani's heritage could be seen as assets to the campaign, so also religion, even as it distinguished Action Council members from one another, could be leveraged to their common cause.

Action Council members did not always agree about how to manage nonlocal community ties. In private conversations, some complained to me that Vijayan's Congress ties introduced political partisanship into the struggle. Others spoke of caste discrimination. Sujit was deeply hurt when his father-in-law died and none of his Action Council friends came to his home to check in on him and express their condolences—a gesture common between friends in Kerala, which he had seen performed on behalf of other Action Council members. In this, and in many other subtle ways, he felt that he and his wife were not always recognized as equals by other local protestors.[14] "Naturally, caste hierarchy won't be plainly seen in the protest," he explained. "But it will be buried deep in there. . . . I can recognize it exactly. I don't know how, maybe just because I've gotten used to it for so long. When I join anything, I'll immediately sense if people are trying to involve me or trying to avoid me."

Complaints like this were always made to me privately; I never heard anyone raise such concerns in Action Council meetings, nor did I overhear these issues contested in the protest tent. Yet in both the complaints and the silences, we can see how bracketing and highlighting drew limits on what issues "fighting for the *nāṭ*" could include. Action Council members' approach to building local solidarity treated other identities tactically, with a light touch; it did not fundamentally challenge the divisions and hierarchies on which these other identities were based. As Sujit's complaint suggests, this approach could perpetuate exclusion—sometimes in ways that were seemingly at odds with ideals of local unity. Insofar as Sujit and others refrained from raising their complaints publicly, the ethic of local belonging, as a public fact, was insulated from such contradictions. Yet it did not make them go away.

Conflicts between solidarity activists and locals often focused on these kinds of contradiction within the Action Council. Vijayan's affiliation with Congress was a constant source of frustration to solidarity activists, who saw it as inconsistent with their vision for people's politics. Even early in the solidarity effort, there were conflicts over the many Congress politicians who visited the protest tent. But even more, solidarity activists saw gender hierarchy in the Action Council as problematic. Dhanya, in particular, saw male dominance in Action Council leadership as a major failing—one she tied closely to her criticism of Manamur activists' lack of "consciousness" and "cause." Women were active on the Action Council; in certain circumstances, as when Sunitha and other women ran for *pañcāyatt* office, they were called upon to be the public face of the protest.[15] Yet most major decisions were made by Vijayan and a few other men—what some called the "core committee" of the Action Council. Dhanya's chief effort at what she called "raising consciousness" (*bōdhavalkarikkuka*) focused on increasing women's leadership roles by organizing the autonomous women's committee. The Action Council members' responses to these efforts—especially, the responses of local women—shed light on the gap between the two groups' approaches to environmental justice.

Several local women were excited about forming an independent women's committee. For example, Rani told me that, while the women's childcare and household responsibilities would make it impossible for them to do sit-ins and hunger strikes, there were other ways that a women's committee could help the campaign. "According to Dhanya," she explained, "If we start a women's committee, then the women's committees [from other people's protests] will come to support us. Then our protest will really be active, if we have everyone's support." Even as she paraphrased Dhanya's arguments for the women's committee, however, Rani's comments reflected a basic difference between her support for this initiative and Dhanya's. Rani did not ground her support in an analysis of gender hierarchy. Rather, like Vijayan's defense of his Congress ties, she focused on the tactical advantages to be gained.

Likewise, when I asked other local women about the women's committee, those who supported the idea talked about its tactical value for the campaign, rather than about a need for gender equality. And when I asked women explicitly about gender hierarchy in the campaign, they consistently expressed that they did not see it as a problem. Some pointed to the prominent roles of several women, such as Rani, Sunitha, and Mary. Others talked about appreciating the work Vijayan and other men were doing and said they did not want to take those roles. Often, like Rani, they explained that women's childcare and household responsibilities made it difficult to participate in some protest activities.

This does not mean that no local women saw gender roles in the campaign as problematic, let alone that they were unconcerned about patriarchy more generally. Consider, for example, a short exchange from my interview with Pushpa, who had held a leadership role in the Action Council in its early years, but who had not been very involved for some time. Initially, when I asked about male dominance in leadership, she seemed to legitimize it.

"Men just know more than us about a lot of things that are going on in the *nāṭ*, or else how to conducted a protest, and that sort of thing," she explained. "So you can't always talk about equal participation."

But when I asked Pushpa, a bit skeptically, how men came to know more about these things, she immediately noted that it was due to the broader gendered distribution of roles in society, not to any innate difference in ability. And she argued forcefully that women were capable of doing anything men did, and doing it better.

"If we just get a chance, then women can do things much more beautifully than a man," she said with a little laugh. "That's no problem. But when it comes to that, there's always that idea that women should always stand behind the men, you know. But just like they run things, just like that, women can do the same."

"Okay," I asked, "but what about the issue of equality then?"

"Oh, you'll never get equality!" she exclaimed, her laughter bubbling over. "But sure, if anyone thinks they can get equality anytime soon, sure let's go with that!"

What makes the idea of gender equality funny to Pushpa? Certainly, it is not that she does not recognize the unfairness of gender hierarchy. On the contrary, for her, demanding gender equality in the Action Council would be laughable because inequality is so entrenched, so integral to the fabric of everyday life in Manamur. It is not that she is not conscious of patriarchy; her consciousness of it makes it seem an impractical cause.

Of course, it is this same entrenchedness of gender inequality in Kerala society that made it a worthy cause for Dhanya and many other solidarity activists. As described in the previous chapter, for Dhanya, the politics of people's protests was a platform for challenging such basic norms and developing a more liberated, egalitarian form of life. The category of "the people" was attractive because it transcended community boundaries, opening the way to transcending boundaries between humans and nature as well. Protest in Manamur offered an opportunity to pursue this cause. Patriarchy within the campaign was, by the same logic, a clear obstacle.

Bracketing party, caste, or religious identity in order to work for "the good of all" may sound similar to solidarity activists' efforts at transcendence—perhaps like a miniature, localized version of their politics of "the people." But in practice, bracketing and highlighting often worked in tandem to fortify the very distinctions and hierarchies that solidarity activists sought to overcome. Bracketing caste made it possible to include both Sujith and Rani in the fight for the *nāṭ*, but it also made it possible to include them differently—one as a painter, the other as a public speaker—and thereby to reinscribe caste hierarchy. Likewise, bracketing gender made it possible for Sunitha, Mary, and Rani to join the fight—even, at times, to stand at the forefront. But setting aside gender distinctions as a basis of organizing, insofar as it worked to focus attention on local belonging, also meant setting aside gender inequality as part of the cause. Even for the most aggrieved Action Council members, their discontent with identity-based bias or discrimination was not the basis for their activism; this was as true for women as it was for men. To the extent that women's leadership could serve the struggle against pollution, they welcomed it. But fundamentally transforming gender relations was not, ultimately, the point.

### *Excluding Nonlocal Motives*

But what if gender equality could be aligned with fighting pollution in Manamur? If Vijayan's ties to Congress, Sanoop's Eid-al-Adha feast, or Sunitha's temple pilgrimage could be turned to tactical advantage, why not an independent women's committee? As Rani noted, Dhanya emphasized such advantages when making her case to local women. More generally, the solidarity activists argued that taking up their vision for an alternative politics—whether by distancing the Manamur protest from political parties or rejecting proposals for moving the factory elsewhere—would only bolster efforts to shut down the factory. Yet, as the case of the women's committee illustrates, such arguments seemed to have little effect on the direction of the Action Council. And in this pattern, we can see that configuring local community—however tactical and opportunistic it could be—also consistently limited the purposes Manamur protestors pursued. Being local was a boundary-drawing process, defined by the causes it excluded.

The boundaries of local belonging in Manamur are clarified by the stories of two men who joined the fight against the gelatin factory on the same afternoon—that afternoon in May when all the shimmering fish corpses came bobbing along the Neelajalam, making their way past villages that the factory's smell had never reached. Many of the people in these villages had not heard about the campaign to shut down the Manamur gelatin factory, but the stench of dead fish drew them out of their homes and into the roads. Stevenson and Rabindranath were among the downstream residents who gathered at a major intersection, blocking traffic in protest of the smell. That day both men learned about the campaign for the first time and decided to become active participants. But in the ensuing months, their stories would diverge.

Rabindranath, who ran a Gandhian commune and retreat center on the shores of the Neelajalam, had longstanding ties to area environmentalists. Initially, he was welcomed by Action Council members for his extensive contacts, while solidarity activists like Dhanya saw him as a key liaison between the newly formed Solidarity Committee and the Action Council. Within weeks of the fish kill, he was visiting Manamur every day, and his involvement was crucial to the intensification of solidarity work from across Kerala. He used his connections with the solidarity activist network to build support for the campaign, but in his interactions with them, he self-identified as a representative of locals. By the time of the Solidarity Convention—which he played a key role in organizing—there were rumors of disagreements between him and Vijayan over organizing process. At an Action Council meeting a few weeks later, even as many of the solidarity activists were beginning to withdraw, Vijayan informed everyone that Rabindranath had been expelled from the campaign.

There were many stories about why Rabindranath was expelled—from accusations of Maoism, to rumors of embezzlement, to his own stories of being falsely accused and villainized. Yet a common thread was his ambiguous, hybrid position as both a local and a solidarity activist. His disagreements with Vijayan stemmed, in part, from this position. In Vijayan's disagreements with solidarity activists—over ties to party politicians, transparency in decision-making, and gender roles—Rabindranath had been sympathetic to the solidarity activists' complaints. Even though Rabindranath's involvement had begun, in part, because of his proximity to the factory, his approach to activism mirrored that of the solidarity activists. He came as someone from downstream, but he also came with a cause.

Such expulsions of "hybrid" activists had happened in Manamur before. When I first went to Manamur in 2010, Narendran and Manu had been central actors in the campaign—part of Vijayan's inner circle. Although both men resided in Manamur, they also had some things in common with the solidarity activists. Narendran, the older of the two, had a long record as an environmental and social activist—he was familiar with *Kēraḷīyam* as well as the small magazines that preceded it. Manu had begun his activism with the youth wing of the Communist Party before coming to focus on environmental justice. By the time I returned in 2012, however, both men were no longer involved in the Action Council and had been ostracized by many of its members. They told a painful story of disagreements with Vijayan, including over decision-making processes and Vijayan's links to Congress politicians. Among Action Council members, rumors of scandal—ranging from embezzlement to illicit sexuality to Maoism—abounded.

The common thread across all three expulsion stories was each man's dual position as someone directly affected by the pollution and as someone working for change beyond his own community. This could be heard in their own stories about conflicts with Vijayan over his undemocratic organizing process and his allegiances with Congress politicians. It could also be heard in complaints among Action Council members about the three men's impractical and overly idealistic approaches to campaign strategy. Most clearly, it could be heard in Vijayan's vague accusations—made to allies who asked why these former leaders were no longer involved—that they had ties to Maoist extremists. Across these accounts, what stands out is the notion that these men were different because they participated in the campaign as part of a broader activist agenda—they were not only there to save the *nāṭ*.

In the eyes of solidarity activists, this difference made Rabindranath an ideal local collaborator. Together with Narendran and Manu, he represented that small part of "the people" that had gained activist consciousness and were oriented toward a broader cause. But for the same reasons that solidarity activists considered these men exemplars of the people, Vijayan and other Action Council members pushed them to the margins. The gap in consciousness between solidarity activists and Action Council members became more than a difference in perspective; it became a boundary dividing the people from the nonpeople, the insiders from the outsiders—and, ultimately, those who had a role in the campaign from those who did not. Rabindranath had been on the wrong side of the line.

This same boundary could also be a reason for inclusion. Such is the story of Stevenson, a Congress politician who, like Rabindranath, lived downstream from Manamur. After the May fish kill, he too began to get involved in the protest. His role was initially more peripheral, and solidarity activists suspected him of attempting to hijack the campaign for his own political ambitions. Unlike Rabindranath, Stevenson presented himself from the beginning as motivated not by environmentalist ideals but simply by his concern about the impact of the pollution on himself and his own community—that is, with person-centric motives like those of the Action Council members. As the solidarity activists' role in the campaign declined, Stevenson's role grew. At the same meeting in which Rabindranath was expelled from the campaign, Stevenson was designated its official *rakṣādhikāri* ("guardian" or "patron") and thenceforth was often in the lead. A man with extensive high-level contacts in the media and government, Stevenson in many ways filled the gap left by solidarity activists' departure. But unlike them, he was one of the locals in both his self-presentation and in the eyes of Manamur protestors—not one of the locals of Manamur, but one fighting for his own people as they fought for theirs.

The Action Council could work together despite differences in caste, religion, and political affiliation, even at times taking advantage of these differences to bolster local belonging. The geographic boundaries of "local" were also flexible, capable of expanding to include those who, like Stevenson, lived far from Manamur. But this flexibility had limits; it depended on Stevenson's person-centric configuration of his motives. He could be local because his motives were configured in a local way. By the same token, when Action Council members drew a line, it was to expel people who identified with solidarity activists like Dhanya and their agenda for radical social and environmental justice. It was when loyalties were divided in this way—between local belonging and commitment to causes that transcended community—that some people were excluded.

These stories of expulsion and inclusion give insight into how the person-centric approach of the Action Council was defined and sustained—how some causes were marginalized as irrelevant to fighting for the *nāṭ*. Why did Sujith refrain from raising his complaints about casteism more publicly? Why did Pushpa find the notion of challenging gender inequality to be so far-fetched? Perhaps because they simply recognized how prevalent and entrenched these hierarchies were, how difficult to change. But perhaps also, in part, because they had seen what happened to those who had tried to use the Action Council to challenge such hierarchies. Perhaps because they saw how Vijayan and other Action Council leaders were drawing the lines.

This came to the fore in the Action Council meeting in which Vijayan brought the women's committee initiative to an abrupt end, which happened shortly after the meeting in which Rabindranath had been expelled and Stevenson inducted. Rani raised the issue, repeating Dhanya's argument about how it could strengthen the protest. Vijayan noted that something similar had been tried before but had proved impractical. He made a vague reference to women having "a lot of other things" to do, tabled the issue, and it never came up again. No one had spoken up to defend Rabindranath in the prior meeting—though privately, several of those present defended him and expressed their regret at his loss. And now, none of the women involved in the initiative, even Rani, spoke up to push for it either. Perhaps they had always felt, in part, that it was unnecessary or impractical. But perhaps also, they did not want to be on the wrong side of the lines being drawn—the lines around the local.

These stories also cast the failure of Dhanya's solidarity efforts in a new light. When Dhanya and other solidarity activists reflected on their disappointment at Action Council members' failure to live up to their ideals for people's protests, they told stories of disillusionment and withdrawal from Manamur. But these same stories can also be told differently—not as stories of voluntarily turning away from the protest in Manamur but as stories of being turned away by protestors there. It is impossible to be sure of the Action Council leaders' intentions in not encouraging locals to attend the solidarity convention, tamping down the women's council initiative, or expelling Rabindranath, but each of these moves undermined solidarity activists' influence and, ultimately, left little room for them to work. Solidarity activists had never claimed to be the main protagonists in people's protests; they reserved that role for the people. But in Manamur, they were no longer given even a supporting role in the politics of people's protests.[16]

## CALIBRATING CARE

Thus far, a closer look at activism in Manamur undermines the contention that local protestors lacked consciousness. Their focus on fighting for their own community was not simply an uncritical reaction to being impacted by pollution; their activism attuned to and amplified the impact. They were not simply unaware of broader concerns like fighting patriarchy or eliminating pollution in every village; they actively marginalized these concerns. But does this not also confirm Dhanya's main criticism? I have called Sunitha's commitment to the *nāṭ* her "cause." But might we not also see this framing of the campaign as support for Dhanya's contention that the Manamur Action Council lacked a cause?

Here, building on Dhanya's use of the term, I take "cause" to mean some further moral purpose in relation to which a goal is important. Is fighting for one's own community really such a purpose? To be sure, saving one's village is more expansive than simply saving oneself, but to the extent that the local protestors' cause has a person-centric structure—prioritizing the welfare of their families, their neighborhoods, their home—it seems to hew more closely to self-interest than Dhanya's vision for environmental justice. Under such a reading, the motives of Action Council members appear to be qualitatively different from those of solidarity activists: the former protect their interests, the latter imagine and pursue a moral purpose.

The pursuit of self-interest has often been seen as orthogonal, or even opposed, to ethics. While Western moral philosophers have long debated whether ethics should be primarily concerned with virtues, obligations, or the consequences of action, they have mostly agreed in contrasting self-interest with answers to the fundamental moral question "How should one live?"[17] The same contrast is present in some environmentalist discourse, which makes impartiality and the maximal extension of moral concern the hallmark of environmental ethics. According to this logic, environmentalisms that make human needs central—let alone the needs of particular human groups—are shallow or incomplete.[18] Likewise, in popular discourse, NIMBYism has often been seen as opposed to environmentalist agendas—think, for example, of local opposition to wind farms and other renewable energy infrastructure.[19] As such, movements that are delimited by locality may also be seen as, at best, an imperfect realization of environmental politics.

Dhanya's criticism of the Manamur Action Council seems to echo this point: it is not just that they have a different cause, but that they do not have any cause whatsoever. By this same logic, one might say that, if they seek only to save their own village from pollution, members of the Manamur Action Council are not real environmentalists. Or likewise: if their motives are person-centric, their activism is nonethical. This echoes the choice that Faiza laid before the art students in this book's introduction between the principled, but uncommon, path of activists' lives and the more prevalent life of the "common people" (*sādhāraṇakkār*), which is defined by greed. Such distinctions were central to how radical environmental activists in Kerala understood their own lives.

Yet, on closer examination, this distinction between interests and values fails to capture much about the activist lives of local protestors in

Manamur. While Action Council members did not deny having a personal stake in shutting down the gelatin factory, they justified the limits of their aims in terms of responsibilities, not interests. For example, Vijayan's close friend and confidant Ashokan said that he did not consider it his job to offer support to other people's protests in other places; that was the job of those who belonged to those places. At the same time, however, Ashokan felt a strong sense of responsibility to work on behalf of fellow villagers who lived on nearby riverbanks and suffered the worst of the pollution. More subtly, Sunitha's sense of responsibility can be heard in her categorization of herself and other female campaign participants as "mothers," suggesting that it is by virtue of this role that she and others were inspired to take action. Like Ashokan's assertion of greater responsibility for members of his own village, Sunitha's invocation of family roles distributes responsibility according to degrees of social connectedness.

Participants in the Action Council practiced responsibility for the *nāṭ* by taking care of one another. They helped one another to share the burdens of protest, pooling money to cover the costs of legal fees and visiting one another in the hospital after clashes with police or factory workers. But mutual care extended further as well. When there was a wedding or a feast, one could expect every Action Council participant to attend. More strikingly, when one protestor's father died, Vijayan and a small cohort of Action Council representatives attended the funeral—an event usually attended only by family and the deceased's closest friends. For Vijayan, especially, such expression of care reached deep into protestors' family lives. When another protestor was struggling with conflict between his family members, he called on Vijayan to mediate. Such roles grew out of Vijayan's position as a leader in the gelatin factory protest. They also strengthened this position, placing him at the center of a tight-knit community.

The limits Action Council members set on their responsibility had to do, in part, with their assessment of the limits of their capabilities. For example, Vijayan expressed an interest in supporting other environmental justice campaigns but said that his heavy involvement in the Manamur campaign rendered him incapable of playing an active role. This is not to say he saw himself as less capable than those, like Dhanya, who contributed to many campaigns. Rather, he was limited by the intensity of his activity in Manamur. During the six months following the police violence, Vijayan's activity became even more constant and frenetic than usual. He hardly slept at all and spent all his waking hours on his phone, on his motorcycle, or both—drumming up support, discussing strategy, or organizing the next event. It was during this period that his health deteriorated most severely; by the time the solidarity activists began to withdraw, he too had been forced to take a step back. Likewise, during this intense period, many Action Council members gave all they could to the campaign, and it exacted a high toll. Ashokan, who talked of limited responsibility for other villages, had not gone to his usual job as a day laborer at construction sites for months, and all his savings were gone. Some protestors' kids were failing classes because of so many evenings in the protest tent. Other protestors were breaking under the psychological stress. Supporting protests elsewhere seemed impossible.

Talk of the limits of responsibility and capability did not mean, however, that Action Council members saw no purpose in their activism beyond stopping the factory's pollution. Though most, like Sunitha, had joined the campaign because of the pollution, some had since begun to imagine and desire greater things. This became particularly apparent when, in the aftermath of the police violence and consequent backlash, it appeared that the factory might actually close. For a few months, the factory halted production, and both locals and solidarity activists speculated that it might never open again. In the midst of such hope, Sunitha told me that she was sure that, even when the factory was gone, the Action Council's work would carry on. She envisioned a transformation to a volunteer group that did service work.

"We can do some little things," she said, "We can't do big, big things. With our organization, our friendship, we can do something for the school, for the kids. We have been planning many things of that sort."

Ashokan, meanwhile, had grander plans. He said that his family and several others were planning to start an agricultural commune based on the relations of reciprocal support that they had developed during the campaign. They would work together to grow their own food, and their land would be jointly owned: "We are giving a new concept for society—some few people buying a bit of land together and sharing it. A lifestyle without walls, like the old way."

Like talk of responsibilities, these plans for a post-campaign future expressed values that are not easily reducible to an analysis of interests. Most obviously, Sunitha's plans included helping others. But more generally, both Sunitha and Ashokan were actively engaged in imagining a better social world and, whatever their initial motives, this vision had become a motive in its own right. Thus, they embarked on a process of imagination and desire that in many ways parallels the process by which solidarity activists like Dhanya and Adarsh arrived in Manamur. And like the solidarity activists' journey, neither Sunitha nor Ashokan knew where they were going when they set out; it was through their participation in the campaign that, gradually, they had come to envision a better life.

Nonetheless, this was not the broader vision of solidarity activists. Sunitha was still focused on serving people in her local area, a limitation she tied to the inability of Action Council members to do "big, big things." Ashokan called his vision a "new concept for society," but it was a vision for communal life among a small group of families, a concept founded on the same principles of belonging and responsibility that had drawn Ashokan and others into the campaign. Both Sunitha and Ashokan looked, with hope, beyond the end of the pollution. But they did not look beyond the *nāṭ*.

In this sense, the local activists' moral purposes aligned with what feminist philosophers have called an ethic of care. Against a valuation of detached impartiality in some moral philosophy, an ethic of care emphasizes the specific roles and relationships of responsibility in which we are, in practice, called upon to care for one another.[20] Joan Tronto writes that we extend and receive care through a "web of relationships," rather from some ideal position outside of social life. This does not preclude responsibility for more socially distant people, such as those others who might suffer if the gelatin factory were moved to their village. But it recognizes that our moral lives are tightly entangled with the lives of others—especially when it comes to the work of taking care of others, as opposed to merely caring about them. Like Sunitha and other mothers in Manamur, we often find ourselves entrusted with special responsibility for particular people. A correlate of such roles of responsibility is the impossibility of any person, or any group of people, caring for all the needs of everyone and everything; like Ashokan, we must sometimes hope that others will take care of their own. In principle, there are no limits to how far care can be extended, and some philosophers have argued for an ethic of care that affirms the extensionist moral aspirations of environmentalism.[21] But here, I find the notion of an ethic of care helpful because it acknowledges that our community ties will, inevitably, shape our efforts to help others, and it calls our attention to the difficult decisions we all must make about which causes to take on.

In this light, the efforts of local activists in Manamur to configure the limits of local community can also be seen as efforts to calibrate the distribution of care. When Action Council members gathered in the protest tent to celebrate a Muslim holiday or hopped on a bus for a tour of Hindu temples, when they attended each other's funerals or failed to check in on

Sujit after his father's death, when they expelled Rabindranath and inducted Stevenson, they were sketching out their own webs of interdependency. They were shaping what it meant to belong to the *nāṭ*. And when they talked about the limits of their responsibility or their aspirations for life together after defeating the factory, they were mapping these contours of belonging onto the aims and purposes of their activism.

## SCALING ENVIRONMENTAL JUSTICE

Compare this sketch, or map, of belonging and care to the more literal map of people's protests found in *Kēraḷīyam* magazine (see map 2 in chapter 1). Like the local activists' talk of saving the *nāṭ*, the map of people's protests offers a picture of a purpose—a perspective on what is valuable or important. In the *Kēraḷīyam* map, every people's protest is a single dot on a flat plane viewed from above. From this perspective, people's protests can look qualitatively equivalent, differing only in their position within this geographical field. The importance of each protest, including the one in Manamur, lies in its contribution to a broader politics of the people and a broader vision for environmental justice. By contrast, local protestors' talk of fighting for the *nāṭ* represents the importance of protest in Manamur differently; the fight against the gelatin factory is important to Manamur activists because it is polluting their village and compromising the health of their families and neighbors. Both of these maps distribute moral concern across geographic and social distance; each gives its own moral bent to social space. They are two different ways of viewing, and enacting, the scale of environmental justice.

Anna Tsing describes a scale as "the spatial dimensionality necessary for a particular kind of view, whether up close or from a distance, microscopic or planetary."[22] In this case, the difference between the scales used by radicals and locals was not just a matter of zooming in or zooming out—of activism in one village or activism in many.[23] Rather, the differences were in how each treated the importance, or value, of activism in Manamur versus activism elsewhere. In other words, the two groups of activists cannot be well-described by circling different areas (one smaller, one larger) on the same map. They were working with fundamentally different kinds of maps.

Timothy Ingold has argued that the politics of environmentalism introduced a similar duality into the meaning of the English term "environment." Previously, he holds, an environment was that which surrounds a thing.[24] Yet with environmentalism, and especially with the notion of global environmental change, the word came to mean a shared object of moral concern: *the* environment—not something we live within so much as something we operate upon. The usual Malayalam gloss for environment, *paristhiti*, shares this dual valence—it can be, simply, a surround (one's yard or neighborhood), yet it is also used in such terms as *pāristhitika pravarttakar*, "environmental activists," to denote the environment of environmentalist discourse. Ingold conceptualizes this duality as one of sphere versus globe. As sphere, an environment is specific to particular people; one perceives it from within as a space that extends outward, in receding layers, away from oneself. As globe, the environment stands apart from all people, from whose perspective it appears as a surface, without the onion-like depth of the sphere, like a map on the page of a magazine. Ingold argues that these are more than two meanings of "environment." They are two ontologies, two views on the shape of existence itself, each of which entails different ways of engaging with the world. In the different perspectives of solidarity activists and Action Council members, we see that these can also be two views of how things have value and importance, two different approaches to how one should live.[25]

Ingold argues that there has been a tendency in Western thought to view distanced, global perspectives as inherently superior to spherical perspectives. This hierarchy can be heard in criticism of environmental justice movements by environmentalists who emphasize, in Ingold's sense, a more global meaning of environment. Such criticism claims that environmental justice is only a partial or limited variety of environmentalism because it focuses on the environments of particular racial groups or social classes, rather than on all humans, or because it focuses on the environments of humans, rather than on all beings.[26] By contrast, insofar as more "global" environmentalisms seem to look down upon the world from above—what Keane calls a "God's-eye view"—they may be felt to perceive higher purposes.[27] Insofar as they seem to survey wide swaths of the moral landscape, they may be felt to reflect more comprehensive understanding—or, in Dhanya's terms, "consciousness" (*bōdham*).[28] Such hierarchies have an intuitive appeal, but the activism of the Manamur Action Council puts them in doubt.

### *Refuting Progress*

Within environmental ethics, arguments for the superiority of morally expansive environmentalisms are often made by linking moral breadth

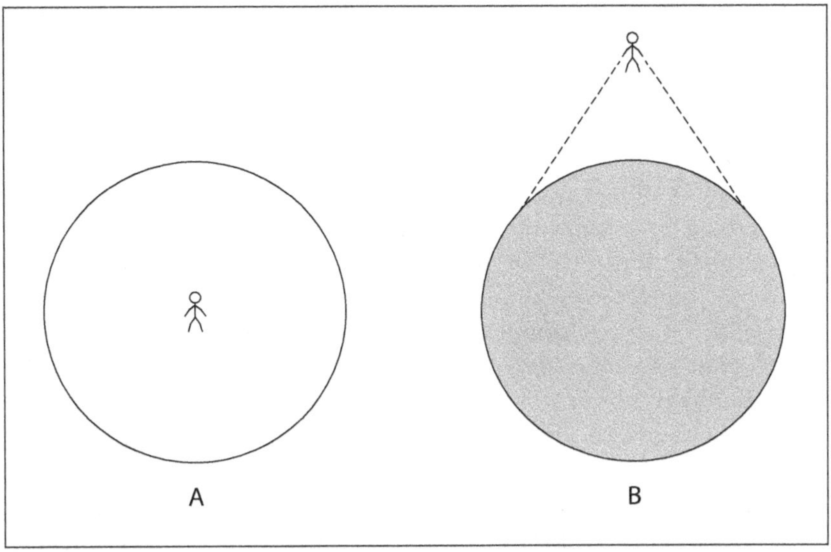


FIGURE 10. "Two views of the environment: (A) as a lifeworld; (B) as a globe." From Tim Ingold, *The Perception of the Environment: Essays on Livelihood, Dwelling and Skill* (2000).

to narratives of moral progress. Historical versions of this argument make environmentalism the endpoint of a process of moral evolution, in which humanity gradually recognizes the inherent value of ever-wider circles of being, advancing from primitive prejudices to enlightened impartiality.[29] In the process of moral "identification" described by Arne Næss, founder of the deep ecology movement, breadth is rendered in biographical time, with individual persons gradually identifying their own good with the good of ever-more-distant others.[30] When people conflate person-centric values with self-interest, as in some NIMBY accusations, they sometimes imply a similar tale of progress. As motives of human action, interests are often regarded as automatic or basic to a person's existence; by contrast, values are motives that people choose and cultivate. In each of these narratives, the alignment of spatial broadening with chronological advancement gives a powerful sense that more expansive moralities are a clear improvement on the narrower, more traditional focus on one's own communities.[31]

Even as stories of progress from narrow to broad can marginalize environmental justice as a primitive or partial environmental ethic, such stories have also been used to describe moral progress within environmental justice movements. For example, McGurty's historical analysis of protests of landfills by the predominantly Black residents of Warren County describes how local protestors initially took a NIMBY approach to activism, but that this "failed because it did not have an elaborative and universal structure."[32] Through collaboration with civil rights leaders and major environmental groups, the local protestors shifted to an environmental racism frame, which "transformed the self-interest position of NIMBY to a broader critique of hazardous waste policy."[33] The environmental racism frame was, in turn, eventually superseded by an environmental justice frame that could include class and other kinds of inequity. Note how closely this account parallels the distinctions and notions of progress that have been used to marginalize environmental justice relative to other varieties of environmentalism.[34] Without refuting the factual accuracy of this account, it is worth considering the predominance of the notion that humans inevitably advance from narrow to broad. Given that the Warren County movement is often seen as the origin of the US environmental justice movement, it is also worth asking how thoroughly integral, how baked-in, these notions may be to environmental justice activism.

The Manamur Action Council offers a different kind of story about environmental justice, one that belies the inevitability of such notions of progress. Local protestors were not primitive environmentalists who had not yet awakened to seeing and doing environmental justice in more global ways; they recognized the connections solidarity activists were trying to draw to broader causes, and they resisted such expansion. Fighting for one's own community was not a default position; it was a position defined in opposition to being an "environmental activist." When they expelled Manamur residents who acted too much like radical environmentalists, or when they inducted the nonresident Stevenson, local protestors moved further in this direction. One might agree with Dhanya that this was not a good thing—that it represented regress rather than progress. Nonetheless, this story refutes the misconception that all person-centric movements—if only they understand—will eventually evolve toward greater impartiality.

### *Inverting Encompassment*

Part of the power of global moral perspectives is that they do not simply conflict with person-centric viewpoints, but also subsume and encompass them. Thus, while solidarity activists criticized locals in Manamur for their lack of more expansive motives, the politics of solidarity also presupposed person-centric motives on behalf of "the people." As noted in chapter 1, solidarity activism depended on the notion that the people were already out there, fighting for their own communities. And even as Dhanya said that locals lacked a cause, some other solidarity activists said that the trouble with the Manamur Action Council was that they were insufficiently desperate to preserve their own lives.

"This is a middle-class natured campaign," complained one solidarity activist. "This is a campaign run by people with money. I don't mean that it's people who have lakhs [hundreds of thousands] of rupees. These are people who go to work every day and save up money and do a campaign. That means: what is not there right now is a situation where people, very seriously, cannot live without ending this. It is a soft campaign."

At the heart of such complaints was the notion that the local activists in Manamur were not really the people at all—because the people were, by definition, those who were fighting only to survive. Perhaps, instead, they were merely the common people (*sādhāraṇakkār*), whose politics were self-interested or, at least, limited to the ethics of community belonging that the radical environmentalist activists sought to overcome. Along the same lines, some solidarity activists assured me that the Action Council would inevitably dissolve and, because the factory would continue polluting, a *real* people's protest would eventually rise up in Manamur. When that happened, they would be ready to return to offer their support.

On the surface, this complaint about the Action Council may seem to clash with Dhanya's complaint that locals were not fighting for a greater cause. Yet solidarity activists' vision for people's politics renders these two complaints compatible. Building on the map image, the scalar view of solidarity might best be depicted as a map of epicenters of local protest, each with its own person-centric structure of motivation, the outer rings intersecting in Venn-diagram style. From this vantage point, person-centric motives are perfectly compatible with commitment to a larger movement, visible in the weave of intersecting rings. It is from the proliferation of person-centric protests that the all-encompassing, community-transcendent cause of the people is born. For solidarity activists, the problem was not that the locals had a person-centric scale; it was that they did not act as if theirs was one person-centric scale among many.

What such accounts of transcendence fail to recognize is that person-centric moralities can also encompass and subsume their map-like

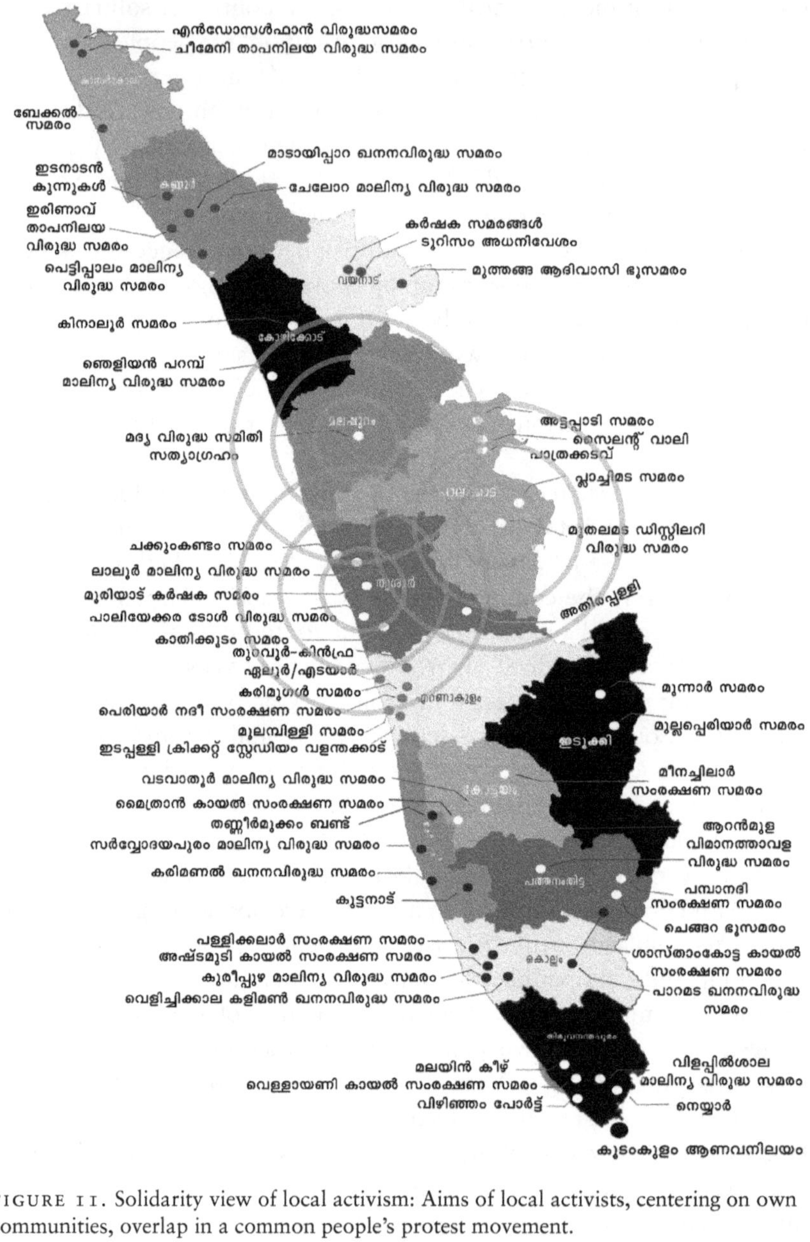


FIGURE 11. Solidarity view of local activism: Aims of local activists, centering on own communities, overlap in a common people's protest movement.

counterparts. Like solidarity activists, some Action Council participants recognized a fundamental difference between the motives of the two groups. For example, when I asked Sunitha why Dhanya and others no longer came to Manamur, she described calling some of the younger activists on the phone to put this very question to them. "Their perspective is different," she said. "They tell me 'We haven't left, Sunithachechi. For us, Manamur is not the only problem. We are involved in every people's protest.'"

Viewed from Manamur, however, the supposedly transcendent "consciousness" of solidarity activists was itself a social position, the perspective of a distinct community. This became especially apparent after the solidarity activists left Manamur. When I asked Manamur protestors how they felt about the withdrawal of outside support, many seemed confused. It was then that I realized that locals had a specific term for people like Dhanya, calling them "environmental activists" (*pāristhitika pravarttakar*), because it was only by using this term that I could make people understand what I meant. Even then, it often took some explication to establish that any withdrawal or departure had, in fact, taken place. To them, the "environmental activists" were not the essential catalysts of people's politics, whose vision unified the Manamur campaign and other local protests in a single, geographically dispersed movement. They were simply one outside group among others—alongside certain parties and politicians, NGOs, or religious groups—each of which might be a source of support for the fight against the gelatin factory.[35] Thus, when they had arrived to hold their solidarity convention, it seemed only natural to assume that it was not an event for locals. And when they left, Vijayan and others professed that they had hardly noticed. Such people were peripheral.

Just as figure 11 represents how the solidarity activists subsumed the Action Council's person-centric scale into a map-like scale, so figure 12 shows how the Action Council's scale subsumed the work of the "environmental activists." Comparing these two ways of seeing and enacting environmental justice, it is apparent that neither is really "broader" than the other. Both are capable of distributing value across an equally wide range of entities—they just distribute value differently. It is not that the map-like view commands a broader view, only that it flattens out the distribution of value across that view. Likewise, each scale is capable of encompassing and subsuming the other scale. And in each case, because of their differences, the other perspective is rendered as warped or obscured. If the map-like scale makes the locals appear self-

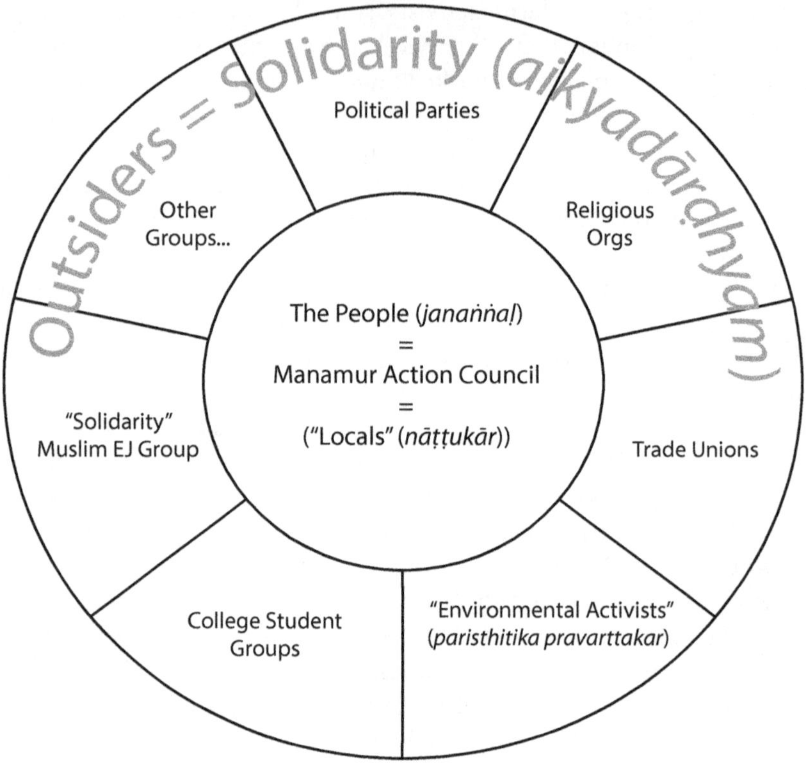


FIGURE 12. Local view of solidarity activists: "Environmental activists" are only one outside group contributing solidarity to protest in Manamur.

interested or short-sighted, the person-centric scale shows the solidarity activists to be insufficiently self-aware or off base.

From the locals' perspective, the problem was not that the so-called environmental activists were fighting for something larger than Manamur; it was that they did not act as if they were simply working within their own perspective, fighting for one agenda among many. Instead, they acted as if their agenda was the only real agenda, subsuming the agendas of all others. Talk of solidarity notwithstanding, they seemed to be elbowing their way into the center of the Action Council's world.

### *Performing Perspective*

Yet these perspectival conflicts should not be too straightforwardly credited for the dissolution of collaboration between the two groups.

Beyond this mutual incorporation of the other perspective, there was also much ambiguity and heterogeneity within each group. Not all solidarity activists were comfortable with talk of "consciousness," and Adarsh chafed at disparagement of the Manamur Action Council's status as the people.[36] Some local activists, meanwhile, admired the solidarity activists' commitment to a larger purpose. As in Sunitha's acknowledgment of their "different" perspective, such admiration still tended to position the "environmental activists" as a distinct and separate group. Nonetheless, she and some others also seemed to recognize and appreciate their mode of activism.

Such ambiguities point to the possibility that, even given such differences in their aims, solidarity activists and locals might have found ways to work together. Yet it is easy to overestimate this possibility. Dhanya's criticism of the "motors inside" local activists notwithstanding, the clash between activists' motives was not simply a matter of differing mindsets. Already, I have shown how Vijayan and other Action Council leaders used processes of inclusion and exclusion to configure the scale of localness, drawing boundaries around what could count as fighting for one's own community. Even if some Action Council members were concerned about issues like casteism and gender inequality, these concerns were difficult to raise because they had been pushed beyond these boundaries. The scalar perspectives from which protestors defined their purposes were not only a matter of what they felt in their hearts; their purposes were constrained by the internal politics of organizing. And even the Action Council leaders had only limited control over the scale of local protest in Manamur. In seeking to position the Action Council as a legitimate manifestation of "the people," they were constrained to respond to other claims within a contentious political arena, where Action Council leaders were far from the loudest or most authoritative claimants.[37]

For example, after the police baton charge, the gelatin company and the police sought to frame the violence in Manamur as instigated by "extremists" (*tīvravādikaḷ*) from outside the village. Solidarity activists demanded that the Manamur Action Council publicly refute these claims, especially after one Congress politician close to Vijayan implied that they were true. But although Vijayan acknowledged the claims were false, he was not as vocal about it as many solidarity activists would have liked. Some of the younger solidarity activists experienced Vijayan's silence as a betrayal. They acknowledged that they were outsiders and proudly avowed being extremists as well, but they felt Vijayan was cooperating with the company's efforts to drive a wedge between them and the Action Council.

This episode highlights the limits of activists' freedom in scaling their own motives. The company's claims about meddling outsiders asserted a scalar perspective on protest of the gelatin factory that circumscribed the representatives of the people to particular locales, while nonlocalized mobilization was rendered "extremist" and targeted for repression.[38] Prior to the police baton charge, the conflict between insider and outsider perspectives had been primarily a conflict over internal aspects of the organizing process, such as gender roles and ties to political parties. Now it also became a conflict over how the campaign would position itself within a broader public discourse of people's protests. This exacerbated already existing tensions within the campaign and put Vijayan in a difficult spot. Any accession to or refutation of the company's claim—or even, as it turned out, any reluctance to refute—would be less a statement of heartfelt conviction than a political tactic vis-à-vis the Action Council's potential allies and opponents.[39]

One might argue that, at least in this instance, this tactical aspect of discourse about motives bolstered Vijayan's control over the meaning of environmental justice in Manamur. While I never got a clear answer from Vijayan about why he did not more publicly refute the company's claims, his silence was not inconsistent with his other words and actions, which had in many cases served to circumscribe "the people" along lines that paralleled the gelatin company's rhetoric. Perhaps this is one reason that Vijayan did not make more effort to oppose these claims: they aligned with his own efforts to sustain the Action Council's person-centric orientation.

Yet Vijayan's silence can also be seen as paralleling Sujit's silence about casteism or Pushpa's silence about gender inequality—that is, as a sign of what had been made difficult to say. Within the larger political arena of people's protest, fighting for one's community was a position the Action Council had taken, a role that was largely scripted by others. If the Action Council aligned too closely with solidarity activists, it risked slipping from that position and losing its legitimacy as the people in the public eye. Thus, even if the company's accusations bolstered Vijayan's influence over the scale of the "cause" in Manamur, they also demonstrated that neither Vijayan nor the solidarity activists fully controlled the terms of their conflicts over the proper motives of environmental justice activism.

## LIVING FOR A CAUSE

What does it mean to live for a cause? Thus far, this book has offered two broad answers. In chapter 1, I showed how radical environmental activists like Faiza, Dhanya, and Adarsh separated two key aspects of social life, *living for* and *living from*, in order to reorganize their lives around the fight for the people and the environment. In this chapter, I have shown how local activists like Sunitha, Vijayan, and Ashokan sought to harmoniously blend these two aspects of social life in order to fight for *their* people and *their* environment. Each of these forms of activist life is integral to organizing for environmental justice, which bridges at least two different meanings of environment.[40]

These two approaches to activism need not always conflict in practice. While the distinction between the people/locals and those who fought in solidarity was present in every environmental justice protest I visited, the tensions between the two groups rose higher in Manamur than elsewhere. Some solidarity activists talked as if the trouble with the Manamur collaboration was unique. One of the most seasoned activists, for example, contrasted Manamur with a recent protest against a dam on the same river, about twenty miles upstream. There, the people had been *adivasis,* members of a marginalized indigenous tribe, who had lacked resources, contacts, and formal education and therefore had looked to solidarity activists to take a lead role.[41] The collaboration there, he noted, had been far smoother. Moreover, he suggested that divisions between locals and outsiders had not been so rigidly drawn in that campaign; everyone had worked together to save the river, which benefited them all.

In Manamur, divisions were rigid in part because Action Council leaders made them rigid. More generally, tensions between differing perspectives were heightened by the relatively balanced power dynamics in the collaboration between solidarity activists and locals. Compared to the campaign to save the waterfall, the locals in Manamur were more mixed in terms of class and caste. Nor were they, as a group, necessarily less educated or less well-resourced than the environmentalists who offered solidarity. Vijayan and other leaders of the Manamur Action Council believed that they could not win their campaign against the gelatin factory without help from people outside the village, and they actively sought such support. Yet they were also relatively well-positioned to set the terms on which support was given.[42] In their own estimation, they did not need the help of "environmental activists" to fight for environmental justice.

However unusual the conflict in Manamur may have been, it brought to the fore tensions that are widely felt, in environmental politics and beyond. The conflict not only highlighted differences in perspective, but led activists like Dhanya to determine which perspective was better and why. From the radical environmental activists' vantage point, local campaigns like the one in Manamur were subsumed by a broader fight for the environment and the people as a whole, making the activism of the Manamur Action Council appear narrow and self-interested. On the other hand, for local activists, the fight for environmental justice was the fight for their village, and the radical environmentalists' vision for change was peripheral to this cause. From this vantage point, the activism of the radicals can seem to presume godlike understanding and agency; they seem to lack critical awareness of their social position.

In this mutual critique, we can see different possibilities for an activist life. The point is not to figure out which way of living for a cause is better. Each approach offers valuable insights into the limitations of the other; each appears better in its own eyes. The challenge is to see the merits of each and, if possible, to find ways of working together with those taking a different approach. This failed in Manamur. But my hope is that this story of how it failed may be useful to those who collaborate across such differences in the future.

3

# Uncommon Subjects

## ACTIVISM ALONE

I had not bumped into Hari for many months. Near the end of my fieldwork, I went looking for him. He was not difficult to find. I called him and found out he was living at his parents' house, he gave me the name of the town, and I got on a train. A few hours later, he met me at the station. And yet it felt like I was searching him out, not just paying a visit, because Hari had never been one of those people I had needed to call on the phone or take a train to meet; he had always just been around. He had been one of the first activists I interviewed, and I had encountered him frequently over the years at protests, seminars, or in the office of *Kēraḷīyam* magazine. We had become friends. I went to find him because I wanted to say goodbye, but also because I wanted to find out what had happened to him.

Hari is one of the most enthusiastic and dedicated activists I have ever met. During my fieldwork, he moved in multiple environmentalist circles, always going wherever the action was. He seemed to spend more time on the trains and buses than in any one place. In my own travels, I came to expect that, whomever I followed and wherever they went, I would cross Hari's path again before long. The last time we had met, he had talked enthusiastically about starting an organic farming commune with Saleem, another activist I had known for a long time. But recently Saleem had told me that the farming commune had disbanded, and that he did not speak to Hari anymore, nor did he know what Hari was doing. No one else could tell me more.

From the train station, Hari and I walked to his family home. At first, it was like old times. He filled every moment with impassioned speech about the principles of ecology and his vision for living in harmony with nature, about what was wrong with everything and how everything ought to be. But when we arrived at his home, he changed. He grew quieter, letting his father and his nine-year-old nephew do most of the talking. His parents had prepared tea, and he offered me some, which was remarkable because he had so often told me how tea poisons the bodies of drinkers and exploits the bodies of plantation laborers. Likewise, when his father suggested that we have nonvegetarian food for dinner, Hari made no protest. He did not partake of the chicken, but if I had only just met him, I would have had no idea that he was avoiding meat in more than the usual Hindu way.[1] I would not have known he was an activist. Later that night, after the others had gone to bed, he shifted back to the subjunctive, and we recorded a four-hour interview of oughts and ideals. But I took note of these silences.

Over the course of my visit, I came to understand that Hari was not sharing his activism with anyone anymore. He spent his days in a low-level government desk job, and he liked his coworkers, but he never talked with them about his former life of protests and clashes with police, his approach to organic farming, or his recent experiments with bringing his diet more in harmony with nature. His family knew his radical views, but he and they both avoided the topic. Time at home was largely spent in his bedroom, downloading and carefully reading the treatises of the nineteenth-century European and American founders of nature cure (*prakṛti cikilsa*). He had not given up on his ideas, opinions, and desires—indeed, the tenets of his environmentalism had arguably become more radical and his dietary experiments more extreme—but he did not speak to anyone about these things. Concerning his absence from the *Kēraḷīyam* social circle, he told me only: "There is no space there for me to speak." Hari had companions in his parents' town—he had people to work with, people to eat with—but as an activist, he was alone.

. . .

How had this happened to Hari? At the "rain camp" described in chapter 1, Sunny told the attendees that they were all people "who can envision things differently, who can stand apart and see differently . . . who intervene differently." Environmental justice activists like Sunny and

Hari embraced this kind of difference. It united them. It was fundamental to reimagining and remaking the world as it was not but should be. But Hari's isolation is a stark example of how these same mechanisms of change could create challenges for activists, not only setting them apart from nonactivists, but also dividing them from one another.

What made the radical environmental activists' ethical project divisive was not merely that their specific set of values differed from those of others. After all, the boundaries of religions, castes, and parties are drawn and fortified by such value differences. In their efforts to overcome such boundaries, activists like Hari resisted tying group belonging to any stable value set or creed. Rather, the radicals' ethical project was unusually divisive because it prioritized values *over* belonging. Being an activist, in this case, meant being willing to part ways with others, even with other activists, in order to live by one's principles. Often, this made it difficult for activists to sustain collaborative projects, like Hari and Saleem's farming commune. At times, it could mean striking out alone.

In this chapter, I take a close look at the social consequences of pursuing environmental justice, including consequences for radicals' relationships with one another. In doing so, I shift focus away from the first two chapters' interest in direct intervention in "people's protests" and toward the everyday bodily practices—such as eating, drinking, walking, breathing, and taking shelter—through which some activists made environmental justice a way of life. These practices were the most radical enactment of the moral logic that opposed values to belonging. They were also flashpoints for activists' disputes with one another, including the disputes that led to Hari's isolation. I compare Hari's story with that of Adarsh, the editor of *Kēraḷīyam* environmental magazine, who on most days split his time between the computer at the back of the magazine's office and the tea shop down the street. Adarsh was similarly passionate about environmental justice but found himself at the center of the activists' social circle, with plenty of space to speak. My aim here, however, is not to recommend one approach over another, but rather to explore how these two activists, in the ways they managed their bodies, reveal contrasting possibilities for navigating conflicts between living consistently with one's principles and sustaining relationships with others.

## BECOMING UNCOMMON

One Sunday night in December, Sunny invited Adarsh, Faiza, and me out for a movie at a new theater near his house. I say "invited," but when Adarsh finally arrived, looking haggard and grim, he told me that he had not really been given any choice. As usual, that month's issue of *Kēraḷīyam* was behind schedule, and Adarsh had wanted to work through the night. But Sunny had pulled rank as the magazine's managing editor, insisting that Adarsh join us.

As soon as we came into the lobby of the movie theater, a strange smile lit up Adarsh's face. The fatigue around his eyes was replaced by eager curiosity. He wandered around slowly, pausing here and there to take it all in. Watching people crowd in for the next show—mostly teenagers in jeans and t-shirts—he told me that, here, it was as if he too were an anthropologist. "It's really a whole different culture," he said.

To my eyes, it was hard to see how this theater was all that different from the one in our town, to which Adarsh and Faiza went regularly. That one was smaller and older, with different snacks, but it was also frequented by teenagers and had air conditioning for at least one of its two screens. But Adarsh said it was all new to him. This was his first visit to a "cinema complex," he explained. Otherwise, he had only ever been to "*sādhāraṇa*" theaters.

My ears perked up at this term, *sādhāraṇa*, "common" or "ordinary," the root word for the term Kerala's radical environmental activists used to distinguish themselves from "common people," or *sādhāraṇakkār.*[2] Here again, Adarsh was using *sādhāraṇa* to make such a distinction, marking his difference from those in the theater lobby, but this time the "common" was on his side of the fence . . . or was it? Because he had only been to common theaters, Adarsh seemed to be saying, he was uncommon. Even his affinity with ordinary things made him radical.

Adarsh roamed around the lobby with his little daughter Tara in his arms. He held her up to touch the ornaments on a huge plastic Christmas tree, and they took turns batting at the inflatable Santa that stood beside it. He seemed to be having fun. But he also said he was uncomfortable—that it was too cold—and after a little while, he began holding his handkerchief over his mouth. He looked over at the large snack counter, with its huge glass case and neon lights, and asked aloud whether they might have tea there.

"No, they only have these corporate drinks, like Coca-Cola and Pepsi," said Maya, Sunny's wife. "It's very bad."

• • •

Living with Faiza and Adarsh, I was often aware of the many ways they found themselves out of place. In some sense, this was a part of what they were up to as activists (*pravarttakar*). It was not simply class or social background that made Adarsh a stranger at the movie theater. He was not uncomfortable because he had never been in air conditioning. He and Faiza did not come to such places because they did not like them, and they did not like them because they were against air conditioning, against huge concrete theaters and malls, and against "corporate drinks." From this standpoint, Adarsh could say that those comfortable in these settings, the common people (*sādhāraṇakkār*), had a whole different culture. But the alterity Adarsh experienced in the theater lobby was not the product of an encounter with an entirely unknown social world. Adarsh was different from those in the theater lobby not because he had never been to such a place before, but because of how he had distanced himself from these sorts of people by avoiding places like this. He was out of place not because he was from another place, but because of how he enacted his opposition to this place.

Activists spoke of such distancing practices as *badal,* or "alternative," a designation that closely parallels Sunny's description of rain camp attendees as "different." As noted in chapter 1, activists described *Kēraḷīyam* and other small-circulation magazines as alternative, contrasting them with "mainstream" (*mukhyadhāra*) publications, such as daily newspapers and the popular weekly magazines put out by the major commercial publishers. Like environmental justice activism more generally, this politics of alternatives had historical roots in post-Communist leftism. *Kēraḷīyam* magazine was alternative not only because it was not corporate, but because it carried on a tradition of using magazines as forums for activist collaboration outside of party politics. Within this tradition, the roots of environmental justice were narrated, in part, as a search for a leftist politics that was not dominated by the Communists. Activists came to intervene in "people's protests," such as the campaign against the Manamur gelatin factory, because they were searching for alternative ways of promoting change.

Yet many of the most salient invocations of the alternative were made with reference to practices that had little direct relation to people's protests. Thus, Adarsh quoted a North Indian environmentalist as saying, "Resistance has two directions. One is protests. The other is alternatives." In this idiom, alternative practices promote environmental justice not by taking to the streets but through many smaller acts of resistance within the flow of everyday life. Prominent examples of such alternative practices could be found on the pages of *Kēraḷīyam:* articles promoting naturopathic medicine, bicycle riding, open software, vegetarianism, and organic agriculture. Many activists produced and sold goods that they called alternative—cotton bags, homemade soaps, organic and vegetarian equivalents of popular packaged foods—to support a particular environmental justice campaign or scrape together an alternative livelihood. Such goods were alternative not only because they could not be found on the shelves of major retailers, but because producing and consuming these products was a way of resisting the exploitation of people and nature in global capitalism. With this logic of alternatives, the pursuit of environmental justice could be extended into seemingly every aspect of social life.

For many practices, there were multiple levels of potential alternativeness. Adarsh avoided Coca-Cola and drank tea, but others avoided tea as well, drinking only *jāppi,* an herbal concoction activists developed for this purpose. Some of the most celebrated alternatives did away with key parts of life altogether. At activist gatherings, people marveled at the precociousness of children whose parents had never sent them to school. Some activists, like Hari, developed dietary regimens as alternatives to any form of medicine, even so-called alternative medicines.

As illustrated by Adarsh's experience of alterity in the movie theater, it was largely through these everyday enactments of alternatives that activists came to feel out of place. Other activities—such as marches, sit-ins, seminars, and rain camps—were more socially circumscribed. One could go to a protest or a seminar over the weekend, and still return to work on Monday. One could write dozens of radical essays for *Kēraḷīyam* magazine, and one's neighbors would never know. But avoiding theaters and soft drinks, eating only fruits, bringing one's own cotton bag to the grocery, or not sending one's children to school—these practices regularly brought activists face-to-face with their differentness, allowing them to feel the distance between themselves and the nonactivist "normals."[3]

The differences that distinguished activists were also acknowledged and felt by those they called ordinary. The moviegoers who fascinated Adarsh probably took no notice of his sniffling or his request for tea, but Adarsh and Faiza's interreligious marriage was a topic of much discussion among our neighbors. Even though the religions of their families were readily apparent from their names (Adarsh being a Hindu and Faiza a Muslim name), neighbors sought to confirm with me that they "really" had an interreligious marriage. It was not clear to me whether such fact-checking indicated approval or disapproval, only that each time a neighbor asked, they were marking Adarsh and Faiza as different. Similarly, though a couple across the street often asked Adarsh and Faiza about their jobs (at *Kēraḷīyam* and a small environmental NGO, respectively), these same neighbors told me privately, laughing, that they could not understand "what kind of jobs these were." More earnestly, several women in the neighborhood expressed concern to me about Adarsh and Faiza's baby daughter, Tara. Once, when Tara was about eighteen months old, Faiza had mentioned to one of our neighbors, a curious and talkative woman with young children of her own, that Tara had a fever. The neighbor had suggested a nearby pediatrician whom her own children had seen, but Faiza told her that they avoided English medicine[4] and that, in fact, they had never taken Tara to a doctor of any kind. Thereafter, conversations with this woman took on a new tone; she began to regularly ask Faiza about Tara's health and weave comments about the virtues of doctors and pills into their small talk. Faiza gave no ground in these encounters. She later told me that she was pleased that her alternative practices had helped this woman see that child health care could be done otherwise. But gradually, the two women chatted with one another less.

### *Values over Belonging*

Activists' most frequent experiences of being "uncommon" were of this small, everyday sort—explaining themselves to neighbors, not finding foods or drinks that suited their environmentalist tastes, getting the sniffles in air-conditioned spaces.[5] But they also felt their distance from common people in a deeper, more definitive way. If becoming an activist was, in part, about cultivating habits that gradually made one out of place, it was also, for many, a choice made once and for all (see the introduction). Preceding and framing everyday experiences of strangeness among neighbors were stories of departure from communities of kin, faith, or party (see chapter 1). Some, like Faiza and Adarsh, had married across religion or caste. Others had become atheists or refused the career path chosen by their parents.

The importance of these stories can be seen in how Adarsh, Faiza, and other activists approached making their own families. For the most part, activists insisted that their children be given maximum freedom to choose their own life paths. Adarsh and Faiza brought Tara with them to seminars and protests and taught her environmentalist songs. But when I asked them what they wanted for their daughter's future, they would only say that this was for Tara to decide. Faiza denied even hoping Tara would become an activist.

And many activists' children did not become activists. Francis, whose departure from his Christian roots I described in chapter 1, made his living selling environmentalist books and doing magic shows that were heavy on lessons about living in harmony with nature. But when both of his daughters chose to attend pharmacy school, he did not object. Occasionally, I heard others say what a shame it was that they had chosen to study allopathic medicine (also known in Kerala as English, modern, or Western medicine) instead of nature cure; after all, their father had published a book on nature cure. When I asked Francis about it, he said that he had heard people talking this way as well—not usually to his face, he said, but still he knew there were rumors. But he refused to acknowledge that he himself was even the least bit disappointed in his daughters. Rather, he pronounced this to be the greatest sign of his success as a father: that his children chose a path different from his own.

"My mother wanted me to become a Christian, but I sure didn't do that, did I?" he told me, his voice brightening with conviction. "No! And I don't insist that my daughters do what I taught them either. There's a liberty in that, right? A freedom, right? . . . It's only correct that my daughters should not study nature cure because, by my calculus, in that there lies a space for freedom and selection. That's why I don't feel any guilt about it."

This approach to parenting exemplifies what it meant for activists to put ethical values over community belonging. In the moral lives of many Malayalis, belonging and values were mutually supportive, particularly within families. One was born as a Muslim, a Christian, a Hindu, or a Communist, and this entailed certain ethical values; these shared values, in turn, were the basis for party membership, marriages, schooling and other modes of group-making. Such, anyway, was the general expectation.[6] Environmental activists sought to eliminate one side of this two-way, mutually reinforcing relationship between values and communities. Shared values could be the basis for establishing new groups and experiencing new forms of belonging, but belonging ought not to be the basis for moral commitments or political activities. What most distinguished environmental activists, in other words, was not their ethics but their meta-ethics—not any specific set of values, but their insistence on this one-way flow of influence from values to relationships.[7]

Concepts from linguistic anthropological studies of evaluative "stance-taking" are helpful for clarifying the stakes in activists' efforts to free values from the social pressures of groups.[8] A key insight in this literature is that any evaluative act, such as self-consciously avoiding English medicine, also serves to define the evaluator's relationships with others. Du Bois's "stance triangle"[9] offers a particularly lucid visualization of this point. When Subject A (such as Faiza) evaluates a particular thing, person, or action (such as English medicine) as good or bad, this also defines a relationship between Subject A and Subject B (such as Faiza's pro-English medicine neighbor), who also has a stance on this matter. Du Bois describes such relationships, illustrated by the subject-subject edge of the triangle, in terms of "alignment" or "disalignment." Both kinds of edge defined by the triangle (subject-object and subject-subject) are intrinsic to all evaluative acts; evaluating a given object necessarily entails alignment or disalignment with others.[10] But radical environmental activists reacted against a perceived overemphasis on intersubjective alignment in Kerala and sought to found their alternative ethics solely (or as much as possible) along the subject-object edge. In other words, they tried to make their ethical positions entirely a matter of their evaluation of rights and wrongs, goods and bads, without regard for the positions of others.

This is not to say that activists did not recognize the importance of the "aligning" aspect of ethics. As described in chapter 1, the community-building activities of activists in the rain camp were all about cultivating alignment and disalignment through evaluation. Expressing opinions on agricultural practices, gender relations, or rain was understood to build moral fellowship (*kūṭṭāyma*) with others. Ethical disalignment, such as Adarsh's experience of alterity at the movie theater, also defined activists' social distance from the mainstream. But although community with other activists was viewed as a valued outcome of activists' positions, it was not supposed to be one's motivation for taking these positions. In their self-understanding, this is what distinguished activist fellowship from the forms of community that defined the ethics of the common people (*sādhāraṇakkār*).

In theory, putting values over belonging did not entail commitment to any specific creed or platform. Activists even, at times, celebrated the difference in the details of their commitments. As suggested by Francis's satisfaction with his daughters' career paths, such disagreements were regarded as expressions of moral freedom and, by extension, indicators of the success of the activists' moral project. Nonetheless, this metaethics was also a mode of *environmentalist* ethics, and this mattered greatly for how activists put "values over belonging" into practice. In pursuit of environmental justice, activists selected certain activities as crucial opportunities for demonstrating their commitment to values over belonging. And while the range of these activities was very broad, a certain subset of acts—those having to do with managing the needs of one's body—generated outsized tension in activists' relationships, both

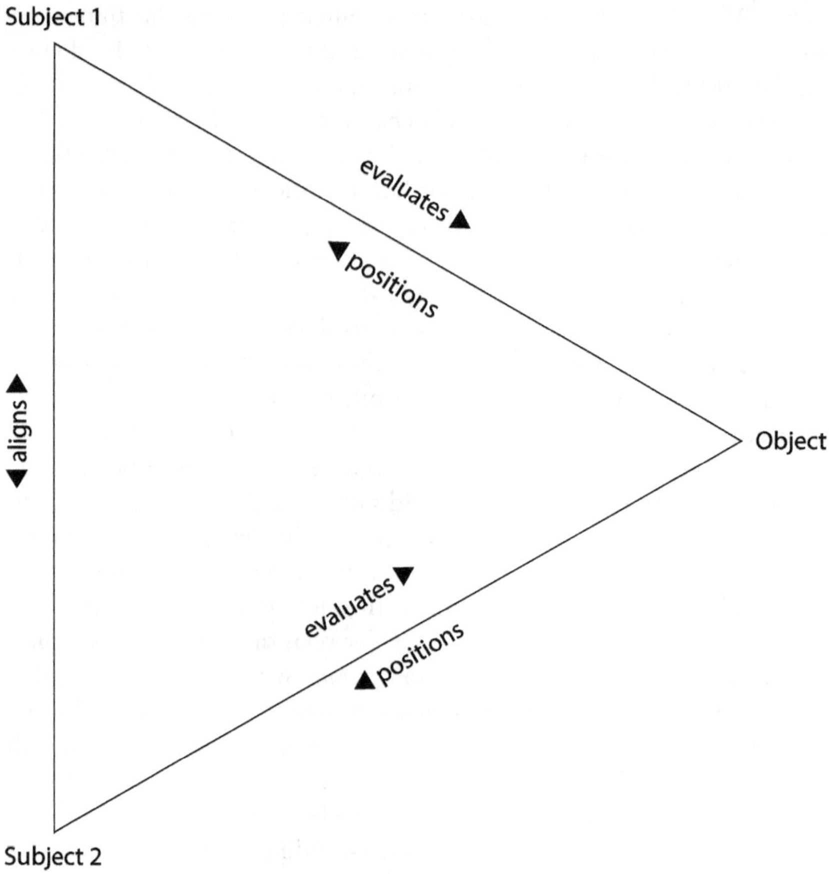


FIGURE 13. DuBois's "stance triangle" shows inherent connection between intersubjective and evaluative relations. Courtesy of John Benjamins Publishing Company.

with nonactivists and with one another. To understand how Hari came to be alone, it is necessary to consider how these seemingly minor, everyday acts, like little perforations in the moral terrain, became fault lines along which activists found themselves divided.

## NATURE LIFE: RESCALING ENVIRONMENTAL JUSTICE

Hari's isolation did not happen all at once. About a year before I went to find him, I saw Hari at the first session of a seminar called One World University—two days of lectures and discussions organized by Mohan-

FIGURE 14. Attendees of "One World University" enjoy a sitar performance during a break.

das, an independent philosopher whom I had met many times in the offices of *Kēraḷīyam* magazine. Hari was one of twenty or so people whom Mohandas selected as having the right "consciousness" (*bōdham*) to contribute to the seminar. As soon as I arrived, I noticed that Hari was a bit quieter than usual. He had been ill, some said, and had been adhering to a fruits-only diet to recover. Nonetheless, he was still very much a part of things, listening attentively during Mohandas's lengthy first lecture, an intellectual history of "alienation" (*anyavalkaraṇam*) in Western philosophy and culture. And when Mohandas called for discussion, he asked Hari to start things off.

As so often, Hari began in the subjunctive mood, pointing out that, since all of those present already had the necessary awareness, the real question to discuss was "what ought to be done." And for this question he immediately proposed an answer. Social change calls for "internal transformation" (*āntarikamāya parivarttanam*), he declared. "There is the question of whether we, each and every individual, are ready. That question is there for me too! The issue is: 'How much can someone do?'"

Hari then used an example that, in hindsight, was an early indicator of the growing distance between himself and the rest of the group. He gave the example of taking a tea break during a discussion about social transformation.

"At one recent camp—at the camp I went to last week, they were also doing just that. A whole lot of this sort of discussion. In great detail, we would discuss all about changing ourselves. But without having a drink of tea, we could not continue with the discussion!"

Grins cut across the edges of the circle, some breaking into titters. Hari pressed on.

"Among some, there is the view that tea and such things are insignificant, while for others, with a more detailed perspective, tea drinking is a very big *teṯṯ*."

This word *teṯṯ* is difficult to translate here—not only because there is no easy English gloss, but because it is unclear how Hari meant it. *Teṯṯ* can sometimes mean something like "error," as in a spelling *teṯṯ*, a minor mistake. But at other times it has a more moral sense, as in confessing a *teṯṯ* to a Catholic priest. Regardless of where Hari's usage stood on this spectrum, he was not looking for laughs. He went on to tie tea plantations to capitalist exploitation of labor and environmental destruction, and he declared that it was insupportable to believe that one can bring about real social transformation while drinking tea.[11] Although Hari made little use of the second person throughout his speech—speaking of "we" or "some people" rather than of "you"—the accusatory implications of this short speech were unavoidable. The seminar participants had just returned from a tea break.

During the remainder of that morning's discussion, the need to do something about tea drinking was affirmed by others several times, though never quite so unequivocally. In closing the discussion, Mohandas also explained that he had requested *jāppi* (the activist tea substitute) for the seminar, but it had not been available. Yet no one seemed to take tea quite as seriously as Hari. As we stood in line for lunch, I heard a couple of other activists teasing Mohandas for "getting in trouble." If the others shared Hari's concern, it was nonetheless not too concerning to laugh about.

## *Nature in the Body*

To understand how activists come to be divided from one another, it is important to understand how and why things like movie theaters, air conditioning, corporate drinks, or tea become so important to them. It is not hard to see the potential for divisiveness in prioritizing one's principles over one's relationships, but this potential was greatly heightened by the expansive scope of activists' moral concern. In previous chapters, I described radical environmental activists' vision for people's protests as a map-like scaling of environmental justice, a detached, bird's-eye view that flattened out distinctions between fighting for one's own people and fighting for others. In radicals' everyday lives—especially, in their approaches to diet and health—such scalar flattening redistributed moral concern, such that acts that were formerly of only minor importance (closer to the "spelling error" meaning of *teṯṯ*) became pivotal acts in the pursuit of environmental justice. The politics of alternatives filled social life with moral urgency—so much so that it spilled over from distinctions between activists and common people, generating division among activists as well.[12]

Hari's call for "internal transformation" as a response to the social issues raised in Mohandas's seminar illustrates how this amplification of moral urgency was accomplished. Hari argued that in order to change larger social and environmental problems associated with capitalism, those present must begin from change within themselves. Hari's problem with tea drinking was grounded in this location of broad social problems within activists' own bodies. Although attendees at an earlier camp had talked a great deal about these same problems, they had failed at internal transformation because they continued to drink tea. For Hari this was an unacceptable contradiction, because the transformation of the self and the transformation of the world were one and the same.[13]

Hari's argument fit well with the themes of One World University. On the morning Hari made his case against tea, Mohandas had described how present-day capitalist politics, family structure, organized religion, modern science, animosity between genders, and separation from nature were all forms of alienation that imposed false divisions on the world. Over multiple seminars, Mohandas went on to propose a different approach to social life, rooted in the ideas of Sree Narayana Guru, an early-twentieth-century Kerala philosopher, Hindu spiritual leader, and anti-casteism activist whose most famous saying is "One Caste, One Religion, One God."[14] Against alienation, Mohandas called for "integration," in which all of the key binaries that define Western thought—mind/body, thought/action, man/woman, self/other, humanity/nature—would be brought into harmonious union. While it is not necessary to detail Mohandas's proposal here, part of the strength of Hari's call for "internal transformation" was that, by refusing any rigid divisions between the internal and the external, or between personal hygiene and global change, it resonated with this call for integration.[15]

Despite how well Hari's speech tied into familiar moral logics, however, people laughed. Perhaps they laughed, at least in part, because these logics *were* so familiar. Hari's arguments against tea-drinking were clearly recognizable as part of a broader approach to environmental justice called "nature life" (*prakṛti jīvitam*) that had recently become popular among activists in Kerala. Nature life expanded on an approach to health and healing called "nature cure" (*prakṛti cikilsa*), a tradition of alternatives to allopathic medicine popularized in India by M.K. Gandhi.[16] Grounded in nineteenth-century American and European opposition to the increasing dominance of professional doctors, the central premise of nature cure is that the human body, through its interactions with its natural environment, has the capacity to heal itself.[17] For Gandhi, the practice of nature cure, like homespun cloth and homemade salt, offered a way of asserting independence from colonial rule.[18] Similarly, in contemporary Kerala, nature life holds out the promise of bio-sovereignty, teaching that all bodily ailments can be cured through such diverse practices as eating vegetarian foods, breathing through the nose rather than the mouth, or drinking sufficient water. But Hari and other practitioners of nature life extended this logic beyond medicinal techniques, making their bodies key battlegrounds in the struggle against the exploitation of labor, the destruction of rainforests, the pollution of rivers, and other environmental justice concerns. In other words, nature life was practiced as a potent, radical mode of the politics of alternatives, injecting this politics into the most ordinary, habitual bodily acts, such as drinking a cup of tea.

Nature life's focus on body and diet was attractive, in part, because it was consistent with some of the tactics of Kerala's environmental justice campaigns. For example, a protest movement against pollution by a Coca-Cola factory in western Kerala called for boycotting the drink. Later, solidarity activists with the same campaign produced and sold a "local" (*taddēśīya,* in this idiom opposed to "corporate") soap, produced by people in the polluted area, the proceeds of which supported local protestors. More broadly, the scalar configuration of nature life aligns closely with that of contemporaneous global movements for ethical consumption, which take acts of purchasing as opportunities (and obligations) to combat geographically distant social and environmental evils.[19] Thus, even activists who did not explicitly identify as practitioners of nature life often engaged in practices that overlapped with its tenets and strategies for change. Hari's speech against tea arguably exploited this overlap, laying out a scalar analysis that aligned with other modes of activism but made individual bodies the fulcrum for environmental justice.

### *Changing My Body, Changing Yours*

Though the integrative and expansive ambitions of nature life were generally admired, many activists also found the practice of nature life irritating. Nature life practitioners ostensibly focused on their own bodies, but they were also constantly making claims on the bodies of others. Sometimes they directly called out other activists, as in Hari's speech about tea. But most often, nature life practitioners' claims upon others' bodies were the subtle, implicit effects of such seemingly unassuming acts as declining an offer of a cookie or enthusiastically describing the joys of a life without sugar or soap.[20] While such acts were ostensibly concerned only with the management of nature life practitioners' own bodies, other activists nonetheless sometimes found them annoying. And while the immediate response to such minor irritations was usually no more than an eyeroll, the most committed nature life practitioners gained a reputation for being irksome. To call someone a *prakṛtijīvi*, "nature creature"—a term that harbored an implicit smirk—was to suggest that they were not only radical, but self-righteous.[21]

The reasons that nature life irritated activists are apparent when one considers the dynamics of stance alignment and disalignment described earlier. By making many formerly ordinary (that is to say, not ethically important) activities the new battleground for environmental justice, nature life also made disalignments pop up in numerous new places—for example, during tea breaks.[22] In keeping with their prioritization of values over belonging (that is, on the subject-object edge over the subject-subject edge of the triangle), radicals purportedly disregarded or even welcomed these ethical disjunctures. The focus of nature life practitioners on their own bodies reflected this prioritization. And on the surface, even to many activists, the divergence between tea drinkers and *jāppi* drinkers could seem minor.

Yet why drink *jāppi* instead of tea? If the future of humanity or of the ecological system depended on these "ordinary" practices, then disalignments took on greater import. Where the stakes of disalignment were so high, they were also more likely to be read as second-order evaluations: one subject (especially, the one with the less radical stance) was now an object of evaluation by another.[23] To drink tea was to betray the cause. To avoid tea was to accuse those who drank. By injecting high stakes into so much of ordinary life, nature life practitioners' management of their own bodies became readable as statements about how everyone should or should not eat, drink, wash, or walk. As such, some activists' enactments of moral freedom were felt to impinge upon the freedoms of others.

Such moral pressures, once applied, could be extremely difficult to shake off. Living among the radical activists, Ahmed and I often talked about our anxieties over the many little things that were suddenly fraught with new moral significance. We were occasionally called to account for eating the wrong thing or breathing the wrong way; more often, we simply found ourselves out of alignment with the moral practices of those around us. We joked about our shared anxiety over "getting caught" when, glancing over our shoulders, we occasionally snuck away for a cold club soda, a lunch in an AC restaurant, or other such less-than-ecologically-sustainable treats. But jokes aside, the pressures we felt when we were around activists overflowed into these private moments as well. Even alone, a soda was no longer "just a soda" because now it was also an act of transgression, something that needed hiding, something for which one might get caught. Even if we did not personally see anything wrong with soda (or tea), its taste was now tainted by our new anxieties.

While such an atmosphere of anxiety may seem inconsistent with activists' explicit emphasis on ethical freedom, the pressures that produced it were neither unanticipated nor entirely unintended. After all, the dynamics that made nature life troublesome for activists' relationships were fundamentally the same as those that distanced activists from the common people. Whether they were nature life practitioners or not, Kerala's environmentalists took up the politics of alternatives not only in order to change themselves but, ultimately, to change their social worlds. Making ethical disalignments apparent generated pressure for change—whether change in nonactivists or in one another. However, nature life intensified the divisiveness of the politics of alternatives by subjecting numerous formerly minor acts to intense moral concern and discipline. In a world so suffused with moral purposes and fraught with moral perils, putting values before belonging could, potentially, make belonging difficult to attain at all.

## STANDING FAST

At the end of the second gathering of One World University, Hari announced that he would not be participating in subsequent meetings of the group. There was too much discussion, he said, and not enough action. After that I did not see him again until, at last, I went to visit him at his parents' house.

There are many ways to tell the story of Hari's separation from the *Kēraḷīyam* crowd. Adarsh suggested that Hari might have simply been part of a more general withdrawal of nature life practitioners from the broader arena of environmental justice activism. The failure of Hari and Saleem's farming commune offers another explanatory narrative, with multiple actors, multiple accusations, and multiple reasons to part ways. But here, I want to focus on a story that Hari himself told me on the evening I went for that last visit. It is his own story about what happened, and it brings out important aspects of how Hari's approach to radical ethical evaluation had changed during the time I had known him. As Hari told it, it is the story of a conflict between the very pulls that defined the ethics of radical environmental activists—between following through on one's principles and folding to pressure from one's communities.

The story begins about a year before the first session of One World University, during a time when Hari was nearly constantly traveling, attending environmentalist camps and seminars focused on various aspects of nature life.[24] At the time, many of those in the *Kēraḷīyam* crowd were also into natural health and medicine. For Hari, this was a time of intense inquiry and experimentation; I remember meeting him at one point during this period and being impressed with the sheer energy and zeal with which he spoke about the potential of nature life to heal, enlighten, and change the world.

Hari met a young woman named Sunathi at one of these nature life camps, and they attended many together, beginning a romantic relationship. A few months after they met, a small but persistent sore appeared on her lower leg, near her ankle. Hari advised her to switch to a simpler diet and to eat only fruit for one meal per day. Sunathi, trusting the advice from the nature cure camps as well as Hari's own expertise, followed these instructions carefully. The sore grew and became two. Hari advised eating only fruit for two meals. The sores continued to multiply and grow. Hari advised that she rest and eat fruit exclusively, and Sunathi followed his advice.

"Finally the wound reached up the whole leg. It was all bluish and seeping," Hari told me. "This wound, if anyone saw it, they would have said that because it had not been treated it had gotten infected, and that the infection had spread to the whole leg. Any average person, who doesn't understand, they would not say it is not healing. Not only that.

When you only eat fruits, your body gets very thin. You lose weight. So people would say your health is declining."

As far as Hari was concerned, however, the seeping wound was not a problem; on the contrary, it was a sign that Sunathi's body was healing itself by expelling whatever toxins were causing her sickness. There was, however, what he called "a social problem."

"The problem is you can't do this in this society. Because if they see a wound that's been seeping for one or two months, they will make some big problem over it, and make her go to an allopathy hospital and diagnose it."

Hari arranged for Sunathi to be taken to a nature cure hospital, where she continued her regimen of fruit and rest. But the problems did not stop there. Some of Hari's friends, particularly in the *Kēraḷīyam* circle, also opposed Sunathi's treatment. He began to receive phone calls urging him to take Sunathi to an allopathic hospital.

"So these people—all of them, even Saleem—all of these people misunderstood and said that it would be some big disaster. . . . Everyone, even Sunny called me. But I had complete faith. But mentally, I had a lot of conflict because everyone was together against me. That was the state of things."

In the end, after almost five months of fruit and bed rest, the wound healed. Not only that, but Hari reported that Sunathi's asthma disappeared and a chronic skin problem cleared up as well.[25] But despite the apparent success of his treatment, Hari's social problems lingered. Soon after Sunathi got better, Hari attended the wedding of Maya and Sunny. There, he said, people made fun of him and made jokes about how he only eats fruit. Not only at that event, but at camps he attended thereafter, even camps about nature cure, people would tease him and make comments about him. "I hear these kind of jokes a lot," he said. "But they're not just jokes. They're another way of saying 'What you are doing is a *teṯṯ*.'"

Here is that ambiguous word again: *teṯṯ*. Only this time it is not Hari directing it at others, but Hari hearing others directing it at him. The ambiguity that this word contains is also present in the whole of Hari's story. One could hear it as a story about medicine—about conflicting answers to purely medical questions about how the body works, what causes disease, or how best to treat a wound. But here, medicine is a profoundly moral matter—not only because allopathic medicine is generally held to be corrupt by environmentalists in Kerala, but because Sunathi's leg, and perhaps more, hangs in the balance. Saleem and

Sunny held Hari responsible for Sunathi's treatment because she had been following his directions, and even when those directions worked, a stigma remained. Hari believed that Sunathi had been healed because he had refused to be swayed from his principles. But it was for this same reason that he found himself increasingly alone.

"I told them precisely and tried to make them understand," he said. "And I stood, and stood, and stood, and at last the wound dried up."

In Hari's own telling, his falling out with other activists was a direct result of his steadfastness as an activist. By maintaining his stance on tea, diet, and medicine, he put himself more and more out of alignment with others, and eventually wound up at the margins of the moral community. He stood, and stood, and stood, and then he found himself standing alone.

A similar story could be told about the failure of the farming commune that Hari had started with Saleem. According to Saleem, the collaboration had come to an impasse over a disagreement about farming methods. Hari had insisted that the farm had to be not only organic, but also "natural," meaning that no amendments, not even cow manure, should be added to the soil. And, of course, we have already examined Hari's condemnation of tea. In each story, Hari's insistence on following through on his principles put him at the margins of the activist community.[26] In the case of Sunathi's leg, while he clearly anticipated that "an average person" would have opposed his treatment, he seemed surprised by the reaction of those who otherwise shared many of his radical positions. These were people who had attended camps with him, and who knew the evils of Western medicine. Yet they did not want him to follow through.

## COMPROMISING FOR COMMUNITY

In his own telling, Hari's voice was already excluded from the *Kēraḷīyam* social scene by the time he gave his speech about tea-drinking. Nonetheless, his words were not without effect. This became apparent to me during an interview several months later, when Adarsh suddenly invoked and affirmed Hari's concerns in connection with a question I had asked about another objectionable bodily practice—the application of VapoRub ointment. I knew that Adarsh thought using VapoRub was bad because he had advised me not to use it myself, explaining it was a petroleum product invented and sold by multinational pharmaceutical corporations. On the other hand, for months on end, I had watched him rub it into his own temples and forehead at least a few times each day. And this, like Hari's disappearance, was a puzzle I wanted to solve before my fieldwork came to a close. When at last I asked him about it directly, he grinned. But rather than talking about VapoRub, he talked about tea.

"It's like people say about the matter of tea. Should we drink tea? . . . It came up in Mohandas's program, didn't it? We have a political stance about that, but I cannot manage to do it."

Adarsh recognized quite clearly that drinking tea was inconsistent with his political views. He pointed out that he had published an article in *Kēraḷīyam* in which the author examined the repugnant politics of tea drinking. Adarsh said he agreed with every word in that article. He even agreed with those who called him a hypocrite for publishing the article without changing his own tea-drinking habits. "But I cannot manage to do it," he explained. "I don't have that much . . . what to say? I don't have the determination. I don't have enough power to say no."

Adarsh drinks more tea than anyone I know, and he drinks it double strong. In the fourteen months that we lived together, he would occasionally tell me that he wanted to cut back, but it never happened. For Adarsh, as for many Malayalis, drinking tea was an integral part of everyday routines of face-to-face interaction. He drank a lot of tea primarily because, awash in the constant flow of activists through the magazine's office, he engaged in a lot of interaction. Whenever someone stopped by, it would not be long before a trip to the nearby tea shop was proposed.

By Adarsh's own estimation, his inability to live out his views on tea-drinking was a moral failure, one he made no attempt to justify. Yet this failure was not entirely at odds with the cause of environmental justice. As described in chapter 1, Adarsh's work as assistant editor of *Kēraḷīyam* was primarily about connecting people, assembling a network of activists who participate in a particular form of political action. Giving up tea might have given Adarsh more time to edit the magazine, but it would have strained the relationships that tea breaks helped to sustain. Just as Hari's speech against tea-drinking entailed a second-order censure of activists who disagreed with his views, so Adarsh's refusal to join his friends for tea breaks, even if he made no explicit appeal to them to stop drinking as well, would have implied disapproval of their own tea habits. Such second-order evaluations would likely have hindered the connective work that was so central to the magazine's mission.

The desire to sustain activist community impinged on Adarsh's ability to follow through on other moral commitments as well. His opposition to home ownership offers a prominent example. As Faiza sometimes teased, Adarsh was "against the home" (*vīṭinetirē*), a phrase that aptly sums up Adarsh's position because it combines the sense of being opposed to a kind of building and, much more broadly, of being opposed to kinship itself. In Malayalam, the most common word for house, *vīṭ*, can also be used to refer to one's family. For example, the surname used by members of one's family is one's "house name" (*vīṭṭupēr*). Adarsh felt strongly that the institution of the "family home"—the construction, beautification, and re-building of which were central to economies of social status in Kerala—was an environmental scourge. Kerala's rainforests were being felled and its mountains ground down to feed the Malayali appetite for larger and newer homes.[27] But Adarsh's opposition to "the home" (*vīṭ*) was also opposition to kinship systems he saw as patriarchal, repressive of children's individuality, and supportive of caste inequities. Just as house and family are iconically linked in Kerala, so Adarsh's opposition to homeownership was simultaneously a stance on buildings and a stance on kinship.

Yet during the latter part of my fieldwork, Faiza and Adarsh began looking to buy a home. Previously, they had always rented, and during my fieldwork we rented a house together that it was beyond their means to continue renting alone. Now that I would be leaving, Faiza's father had suggested they consider buying instead of renting, and she had begun inquiring about homes in the area. Adarsh refused to be directly involved in this; whenever I asked about it, he said that he had no interest in the matter. But he knew Faiza and her father were going to look at houses and, other than occasional grumbling and complaints, he made no effort to stop them. He was resigned to becoming a homeowner.

"If [Faiza] is interested, then maybe that isn't the sort of thing I should insist on," he said. "Because all of the reasons behind my decisions might be very different. I don't have that sort of fascism [Eng] that she has to always follow my path."

Although Adarsh's compromise on homeownership diverged from his official position on the matter, it was also, in a subtler sense, consistent with the principles that underpinned this position. Elaborating on his aversion to "fascism" in his relationship with Faiza, Adarsh pointed out that the capacity of married activists to live out their principles was highly gendered. If the husband began going to environmentalist camps or eating only natural foods then his wife and children often did so as well. But if the wife took an environmentalist turn, she could not expect her husband to follow her lead. Indeed, most wives I knew in the

*Kēraḷīyam* circle had either accompanied their husbands into activism or, if they were activists before marriage, had married men who shared their views. There were few wives whose husbands were not also activists, and even fewer whose husbands seemed to have followed their wives into activism. By not insisting on avoiding homeownership, Adarsh believed he was resisting this patriarchal flow of values within activist families—a position consistent not only with his opposition to patriarchy but, more generally, with a broader discomfort with the intersubjective moral pressure that pervaded the politics of alternatives.

### *Reconsidering Family*

Adarsh's compromise with Faiza on homeownership was different in many ways from his failure to give up tea. Whereas he made no attempt to justify his tea habit, he had a clear, explicit rationale for not following through on his opposition to family homes. Yet both moral "failures" enabled Adarsh to produce and sustain valued forms of community. Tea-drinking supported fellowship (*kūṭṭāyma*) with other activists, and homeownership facilitated a less patriarchal marriage. Reflecting on his struggles with tea, he suggested that following through on one's radical positions too consistently could have the opposite effect. He saw this happening among some activists who pursued nature life, who (he said) took their views to such extremes that they would not even talk to those who differed from them.

"I disagree with the higher puritan sort of stance in that. Won't talk to a person who drinks tea! I'm against that kind of puritanism [Eng]. That's a terribly . . . in that there's a terrible sort of fascism [Eng]! A prejudice is there, in speaking like that."

Here, again, Adarsh used the term "fascism," and paired it with "puritanism," to challenge the notion that following through on one's values is always the best approach. He went on to talk about how Hari and people like him were not very active in the *Kēraḷīyam* social circle anymore, saying that they had mostly split off and formed their own group. As we have seen, Hari's story was more complicated than that. But what Adarsh keenly identified here was the connection between strident moral conviction and social division among activists. For Adarsh, such division was more than an unfortunate side effect of a radical way of life; here, at least, he placed the onus of responsibility on those who carry their commitments too far. If they lost their friends, it was only because of their own puritanism.

As much as he was aware of his own moral failures, Adarsh was far more deeply concerned about this tendency of activists to split with one another over their differences. In attempting to articulate some resolution to this problem, he suggested, paradoxically, that activists should take family relations as a model. For all his opposition to the home (*vīṭ*), Adarsh admired one thing about mainstream Malayali kinship: compared to other relations, blood relatives did not part ways so easily.

"Take me and my father—we are always fighting, always arguing. If he was not my father, maybe we'd just stop talking to each other. But maybe the reason that relation continues, even though we fight, maybe it's because of that blood relation. . . . But we should have that toward everyone, hear! But that sort of bond, it amazes me sometimes. Even if we have some difference of opinion, that bond still stays! I only see that kind of bond in the family."

At first glance, Adarsh's praise for the family may seem to contradict not only his own opposition to the home, but also, more fundamentally, the meta-ethical principle that I have suggested defines radical environmentalist ethics in Kerala: that values come before belonging. However, what Adarsh admired about blood relations was not their role in producing moral alignment. On the contrary, what he admired was that family members can remain close "even though we fight." In other words, family relations are a good model because they are so unaffected by moral disalignments. And the antithesis of this can be seen among those nature life activists whom Adarsh called "puritanical," who allegedly will not even speak to those with whom they disagree. Among such activists, he suggested, moral disalignment breeds enmity.[28]

Thus, as imagined by Adarsh, making activist fellowship more like kinship is entirely in line with an insistence on putting values over belonging. Adarsh ostensibly wished to eradicate not only the influence of group belonging upon ethical values but also (working in reverse), the potentially corrosive influence of differing ethical values upon social bonds. This perspective could be seen as even more radical than that of Hari and other nature life activists, who rigorously resisted the prior vector of influence, but seemed to accept the reverse as a matter of course. By seeking to counter the alleged puritanism of the nature life activists, Adarsh laid out a vision of morality that was even more purified of social entanglements.

Despite his invocation of his relation to his father, however, Adarsh's vision for activist "kinship" is belied by how he conducts his own relationships with kin, particularly with Faiza. Maintaining family bonds has required that Adarsh not only continue relationship despite differences, but also compromise on some of his commitments, such as his opposition to homeownership. Such compromises could be seen in his relationship with his father as well, who brought Tara mass-produced plastic toys whenever he visited. If relations between kin really were so impervious to moral differences, such compromises would not be necessary. If he could isolate his ethical alignments and disalignments from other aspects of his relationships, Adarsh would not have to choose between following through on his principles and sustaining community with others. In this sense, one can see how Adarsh's compromises on tea and homeownership not only conflict with his ethical values but, more fundamentally, run counter to his efforts to keeping ethics and group belonging separate.

## LIVING IN THE OUGHT

Hari's story of standing fast was unusual. During the period of my fieldwork, I met only a few other activists who were so significantly marginalized for taking "alternatives" too far, and no one else who came to embrace isolation as Hari did. However, the kind of ethical disalignments that drove Hari back to his parents' house also troubled activists' efforts at community more generally. Just as Hari and Saleem's farming commune failed because of disagreements over correct cultivation practices, there were many stories of other failed collaborations, other communes started with great expectation and dissolved in frustration over differences of opinion. Radical environmental activists often talked to me about how much more they could have done if only they could have worked together.

The problem, I heard again and again, was "ego." Radicals generally attributed their inability to overcome their differences to an excess of self-importance and pride, and this may have been a factor in some cases. However, studying stories like Hari's closely, it is clear that what drove these activists apart was not so much self-importance as the importance they gave to rigorously living for a cause. In other words, activists' collaborations were predisposed to fail by the fact that they were made up of activists. Like Hari, these were people who had set themselves apart by choosing to live according to their radical evaluative positions despite the difficulties those positions created for relationships. When they broke with their families or their villages, they chose consistency with their views over continuity of relationship. That is what made them out of place among "common people" (*sādhāraṇakkār*). It is what made them activists. But when they encountered differences of opinion in their own collaborations, their commitment to living as activists often drove them apart from one another as well.

Viewed from one angle, radical environmental activists' willingness to part ways is what made their ethics an enactment of freedom. Refusing to be swayed by demands from kin, caste, or party—or by their own desire to belong to such groups—they instead sought to build communities that were based on a shared vision for change. One might well expect, as many Kerala activists did, that friendships founded on a common cause would be stronger and more enduring than those comparatively arbitrary forms of affiliation into which people are born.

But, in practice, activists' relationships with one another were at least as fragile, if not more so, than these other ties. In the politics of alternatives—especially nature life—activists' enactment of moral freedom put intense pressure not only on people's values and actions, but also on their relationships. Because activist fellowship (*kūṭṭāyma*) was founded almost exclusively on shared ethical views (or, in other words, on alignments of moral stance), it was also that much more susceptible to the rifts that ethical disalignments can produce. Thus, Adarsh observed how fragile activists' ties with one another seemed when compared to the bonds of kin.

Radical environmental activists' own analyses of social life offered resources for understanding this social fragility. In Mohandas's One World University, alienation was presented as a central problem for modern Malayali society, and this alienation was said to spring from dichotomies that parallel the dichotomy between values and relationships. Likewise, in the rain camp described in chapter 1, activists sought to overcome divisions, singing of being one with each other and enacting harmony with nature by sitting out in the rain. But these ideas were not enough to overcome the moral fractures produced by projects of change. On the contrary, the lives of radical environmental activists show how even efforts to break down dichotomies and divisions can divide people from one another.

Adarsh's willingness to bend on homeownership suggests one way of countering this tendency toward fragmentation, especially within activist communities. But such compromises run against the grain of how activists understand themselves and could, potentially, undermine their ability to make change. Sticking to one's principles is not only an enactment of freedom; it is also a means of influencing the ethical lives of others.

This is obvious in Hari's story of standing fast, in which he stuck to his principles by guiding Sunathi in rigorously carrying out those principles. But it can also be seen, more subtly, in Faiza's satisfaction that, by being willing to be seen as strange by her English-medicine-practicing neighbor, she also introduced that neighbor to a different way of approaching health and childcare. If all disalignments were avoided, there would be no alternatives. Without alternatives, there would be no change.

At the same time, even if projects of social transformation are defined by disalignment, they also need relationships. Hari's predicament illustrates this well. Hari had an unusually difficult time maintaining relationships with other activists because he so insistently called them to account. Thus, even though he had already experienced marginalization due to his radical adherence to nature life, he used his chance to lead a discussion as an opportunity to chide them all about drinking tea. By contrast, he put little to no pressure on his parents or co-workers, keeping silent about his views—as if, perhaps, he was less concerned about influencing their views. Yet because he was so adamant and outspoken in his interactions with other activists, he jeopardized his relationships to the point that he had little opportunity to influence them either. The same quandary can be found in other contexts. If Faiza no longer talks with her neighbor, she will no longer be able to introduce her to alternative ways of living.

Activists' desire to change the lives of others thus limited the scope of environmental justice as a practice of ethical freedom. In comparison to the moral lives of nonactivists, environmental justice activists perceived their efforts to challenge norms and live out alternatives as free from the intersubjective side of ethics. They arrived at their values through the modes of activist inquiry described in chapter 1, not by virtue of the groups to which they belonged. Yet making change was inherently intersubjective; without relationships that were based on more than shared ethical values, activists could not make change. They could not reach out to those with whom they differed, nor could they collaborate with one another. Thus, activists could not ground their social lives entirely on their commitment to change; they also had to find ways of maintaining ties that were based on more than shared principles.

This balancing act was not unique to this particular set of activists. Their efforts to purge their ethical lives of social pressure made tensions between purpose and belonging highly visible, but the roots of these tensions lie in the very structure of ethical evaluation. This is not to say that these deep roots always give rise to the same dilemmas. In the next chapter, for example, I describe how members of the Manamur gelatin factory campaign (whom the environmentalists described in this chapter criticized for limiting their concern to their own circles of belonging—see chapter 2) also found themselves at odds with families, friends, and neighbors over their activism, but channeled the divisive side of activist ethics to very different ends. What the stories of Adarsh and Hari make particularly apparent, however, is that there is no escaping this tension—no severing the evaluative edge of ethics from the intersubjective edge; at least, not for activists. Those who set out to transform their social worlds must necessarily remain participants in those worlds even as they set themselves apart. Even the most avowedly liberated moral lives, if they would be activist lives, depend upon the lives of others.

4

# Unquiet Objects

## FIGHTING FOR SYMPATHY

By early afternoon, only a couple dozen protesters remained on the bridge, chatting in clumps of three or four. Laughing boys were still bringing fish up from the canal that flowed behind the gelatin factory and down to the river, laying them side by side on the low bridge wall. The stench swelled. The row of milky-eyed victims lengthened. But the TV cameras had long since come and gone, and the crowd was dissipating. The action was over, or so it appeared, until a little white car pulled up—the sort driven by politicians and journalists. The crowd began to regather.

Initially, the protestors seemed eager to show the reporter and his cameraman the floating fish bodies and tell the story of their discovery. But then someone noticed they were from Kerala News. A couple of months before, after another "fish kill" (a sudden death of the fish in a water body), Kerala News had reported that motor oil and diesel fuel had been found in the river, seemingly confirming the gelatin factory's claims (repeated after every fish kill) that local protestors had intentionally poisoned the water. As the cameraman picked his way down to the bank of the canal, a few heckles could be heard.

"Aren't these the same guys who came that night and then put out that fake news! This is a scam! These guys are frauds!"

"Why did you say that about motor oil and diesel! From now on Kerala News shouldn't come here!"

The cameraman went about his business quietly, making no reply as several protestors followed him down to the river and back, tossing out comments and insults from a few paces away. But the reporter on the bridge above was not so unflappable. My own video camera did not catch what he said, but in only seconds a dozen people had converged in a ring around him. Those closest were shouting. Bodies jostled, fists lifted high, fingers shook in the reporter's face. No blows had fallen yet, but the reporter was matching shouts with shouts, and the ring of bodies was cinching in.

Some tried to calm each other down.

"My friend," said one protestor to another, "let me just say one thing: rage (*vikāram*) isn't wanted here. Don't get overexcited. My own son works for Kerala News."

Another man gave the reporter an opportunity to distance his own reporting from the coverage by his channel, suggesting that "pressures from above" the reporter were the real problem.

"You might be recording the truth here for your report," he explained, all the while trying to stop an older man from jabbing the reporter with an umbrella, "but that's not what's coming out [on the TV]."

But other protestors refused to be calmed, and the reporter was in no mood to back down.

"You have a lot of nasty people in your Action Council!" he shouted. "First you need to throw them out, otherwise your campaign is never going to win!"

The following morning, Vijayan—the Action Council's lead organizer—called me to ask about the video my research assistant, Ahmed, had taken of the quarrel on the bridge. The combative reporter had complained to his superiors, and Vijayan hoped he could use our footage to refute these complaints. I told Vijayan that he was welcome to use what we had, and he said he would come right over. What I did not tell Vijayan was that Ahmed and I were far from confident that our footage would satisfy his needs. The reporter's reaction may have stoked the flames, but it seemed to me that the protestors had started things and that the footage would only confirm that they had been the main aggressors. Besides, Ahmed's footage ended with one of the larger protestors charging at the camera, one palm brandished high over his head, yelling at Ahmed to stop filming. Ahmed had clearly been shaken by this, and we had returned home soon after. When Vijayan called, I had been wondering whether we should delete the video, while Ahmed had been wondering whether he felt safe going back to Manamur. Thus, an hour later, as Vijayan took a seat at Ahmed's desk and began watching this footage, Ahmed and I retreated to another room, anxious that Vijayan might not like what he saw.

. . .

Members of the Manamur Action Council could feel the pollution from the gelatin factory seeping into their bodies, their homes, their lives. It was the film on the surface of the water in their wells, the slipperiness that made rice slower to cook and clothing more resistant to laundering. It was the itchy rash on the legs of those foolish enough to wade in the river. Above all, it was the stench that descended over everything, usually at night, making old men cough and babies cry. This saturation was fundamental to how they understood themselves. They protested, they said, because they could not bear it anymore.

But even as the smell of injustice burned in their nostrils, Action Council members necessarily saw their campaign from other, less intimate angles as well. They saw their own activism through the lenses of TV cameras, the headlines of newspapers, and the tallied likes and shares of Facebook posts. For most of Kerala, people's protests were primarily experienced as strings of events on TV screens and newspaper pages. The rest of the state knew Manamur only through intermittent headlines about dead fish in the river, footage of police ducking stones behind plexiglass shields and whacking backs with batons, soundbites from speeches made by persons of renown, and images of leaders stretched on cots to fast until death. And even as they participated in such events, the Manamur protesters saw themselves from this angle as well, especially via Kerala's dozen Malayalam news channels.[1] Participants in the protest watched the news closely, celebrated increases in coverage, and strategized to bring TV reporters back to the village.[2]

The central challenge was to make these two perspectives on the campaign align. As described in chapter 2, the Action Council members were not concerned with living up to the expansive vision for environmental justice espoused by activists from outside their village. Their aims were tightly bound by circles of belonging. Nonetheless, they believed that the fate of their campaign rested entirely on how they were seen by those beyond these circles. How could those watching TV be made to recognize the worthiness of the Action Council's cause? How would they see the slime, hear the coughing, or even smell the stench?

This chapter explores how members of the Manamur Action Council attempted to convey their injustice-saturated reality to distant audiences. This problem is common in environmental justice movements, though it can appear in many guises. The historian Ramachandra Guha presents a classic example in his study of the renowned *Chipko* (Hindi "hugging") movement in northern India, in which forest dwellers famously halted government lumber harvesting by hugging the trees.[3] Popular accounts of the struggle had depicted it as a spontaneous uprising, in which tribal women wrapped their bodies around the trees in a passionate, "primal" response to an ecologically destructive modern society. Guha argues that these accounts are historically inaccurate; the historical tree huggers were mostly men, their actions were carefully planned, and they were motivated largely by local traditions of forest rights, not by primal environmentalist instincts. Yet Guha also recounts how the Chipko myth, so to speak, contributed to the movement's fame and ultimate success—and how its leaders have since cultivated a public image that plays into the myth, eventually positioning themselves as standard-bearers for Indian environmentalism. This image erased the work that went into making the Chipko movement, but it was a picture of environmental protest that could garner public sympathy.

Decades later, the Manamur Action Council's work benefited from the repertoires of environmental struggle pioneered by movements like Chipko. The Action Council employed two sets of performative tactics, each aimed at stirring the sympathies of audiences beyond Manamur. The first set focused on displays of evidence, both via sensory encounters with pollution and through scientific research. By giving tours of pollution in the village, Action Council members attempted to draw visitors—especially civil society figures, politicians, and government officials—into their own everyday experiences of injustice. By producing research, they sought to tie this reality to discourses of scientific objectivity employed by government agencies and the courts.

The second set of tactics focused on performances of bodily vulnerability, displays of both suffering and uncontrollable rage. Conducted in public streets, such displays sought to garner attention both from phone-bearing, Facebook-linked passersby and from journalists sent by major newspapers and TV networks. Via social media and mass media, the protestors attempted to appeal to the sympathies of the Kerala public.

Ultimately, all of these tactics shared a common logic: protestors obscured their own agency as moral and political subjects, seeking instead to present a sympathetic object to others who could judge and act on their behalf. Here, I use the term "object" simply in the sense of that which is evaluated—the same sense I employed in the discussion of subjects and objects of evaluation in the previous chapter.[4] Subjects evaluate: they announce their views and seek to remake the world according to their moral vision. For radical environmentalists like Hari or Faiza, environmental justice was largely about becoming a certain kind of subject. But for protestors in Manamur, the task was not to evaluate, but to make oneself available to be evaluated in the desired way—to become a certain kind of moral object, "the people," whom others would support. This was not a mute politics, but it meant obscuring their ethical subjectivity—the constant evaluative and strategic work of being activists—so that the injustice in Manamur could be made to speak through their protest.[5]

This performative strategy may seem counterintuitive. Even though humans are both subjects and objects, it is common to see assertions of human subjectivity as ennobling and politically empowering, whereas objecthood is associated with domination and dehumanization.[6] Thus, *objectification* refers to degrading representations or evaluations of others. And for local activists in Manamur, performing objecthood required compromises that, at times, constrained options for self-representation. Moreover, it openly put the fate of their protest in the hands of others, including skeptical government officials, inscrutable politicians, and a distant and fickle public. Yet being this particular object, the people, also gave them powers that they considered crucial to their cause.

## THE SMELL OF INJUSTICE

For the Manamur Action Council members, the day the gelatin factory's effluent pipe broke was a day for outrage. But it was also a festive day. The familiar biting, rotten smell of the pollution hung over the road, but a victorious, even gleeful mood was in the air as well. Peters, an accountant and local civic luminary who was active in the anti-factory Manamur Action Council, stood knee-deep in the muck for several hours while a forest of flags grew up around him, each bearing the insignia of a different political party or union. TV camera operators shifted their tripods from place to place, struggling to get shots of the black gunk and yellowish foam for the evening news. Vijayan, the Action Council leader, zig-zagged through it all like a bumblebee, wearing a wide-brimmed cowboy hat and a gleaming smile. By noon, Peters became faint and had to be carried away, but this seemed only to lift the protestors' spirits higher. At last, the muck that had been seeping through the groundwater

FIGURE 15. Peters stands in the polluted rice paddy beside an open coffin.

into their wells was out in the open for all to see. The miasma that swept through the village at night was even sharper and more pungent beneath the noon sun. This was a festival of evidence.

But the celebration did not last long. In the early afternoon, the District Collector finally arrived.[7] Vijayan had called her office in the morning, and both the protestors and the media had been awaiting her. As she stepped out of her little white car, everyone in the road gathered around, with several prominent Action Council women in the lead. They guided her along the path of the effluent—down the edge of the paddy field to where it emptied into the canal, then along the winding canal to the river—stopping frequently to explain what she was seeing. The Collector said very little, but nodded, smiled, and looked very concerned. Then, when she had seen all and was about to get back into her car, a reporter pointed a camera in her face and asked for a statement about what she had just seen.

"In a layman's way, I can say the problem is very clearly that there is pollution. The government has to decide whether they want a technical report and what must be done."

With that, she slipped back into her car and left. Action Council members continued to linger by the edge of the paddy field, but their mood soured. The Collector had seen the pollution with her own eyes, breathed it and smelled it, and even told the TV reporters that it was real. Yet all that would come of it was another government study.

. . .

Environmental justice campaigns in Kerala, as in other places, put great effort into producing evidence of injustice. In Manamur, the Action Council pursued a wide range of strategies to produce evidence regarding the type and quantity of toxic material coming out of the factory, its presence in the air and groundwater, and its effects on human health. Yet their efforts were often frustrated by a paradox: they lived their lives immersed in pollution; they smelled it in the air, saw it floating in the river or the paddy field, and tasted it in their drinking water. Yet they struggled to translate this experience into forms of evidence that could be persuasive to others. Especially, they had little success in producing the "technical" forms of evidence demanded by government officials or the courts.

To fully understand this challenge, it will help to look more closely at those evidentiary practices that did often succeed. While the Manamur Action Council actively sought opportunities to engage in scientific studies of conditions in the village, this was not their primary approach to evidence. Rather, as in the Collector's visit, producing evidence was first and foremost an interpersonal activity, a guided tour of the sights and smells that they encountered in their daily lives. Through such interactions, local activists sought to bring outsiders inside their sensory world and, thereby, make them feel how intolerable that world had become.

### *A Shared Moral Reality*

The members of the Action Council pressed close to the Collector as they led her through the sights and smells of pollution, enfolding her in a tight envelope of speech and gesture. In my video recording, multiple voices are constantly speaking at once. All that can be made out is a series of pleas to witness, phrases led by deictics[8] such as "this" and "here":

"*This* is the water that we have to drink! We should make them drink *this!*"

"*This* is my riverbank. I am living right next to the river. . . . The riverbank next to my house is like *this*."

"*This* soil cannot be used ever again! *This* soil is destroyed. *This* area is destroyed."

FIGURE 16. Action Council members guide the District Collector along the edge of the polluted rice paddy.

The protestors directed the Collector's attention to aspects of what lay around her. As they pointed with their words, they also pointed with their hands. They leaned forward to insert themselves into her visual field. Recounting stories from their own lives and the lives of their families and neighbors, they tied what the Collector saw to aspects of their experience that she could not see—to drinking the water, to living beside the river, to farming the soil.

When locals guided the Collector's attention to the foam on the water or a house on the far side of the paddy, they sought to construct a shared, intersubjective orientation to certain objects. For the sociologist Alfred Schutz, such intersubjective coordination to objects is what allows multiple people, each with their own perspective, to inhabit a common social world.[9] In displaying the pollution to the Collector, Action Council members sought to bring her into their moral world. This was a thoroughly value-laden process—not only an epistemological coordination, but also a moral coordination.[10]

In chapter 3, I introduced Du Bois's "stance triangle" to show how the moral evaluation of common objects, such as tea or air conditioning, could impact activists' relationships with one another.[11] When activists took a stand for or against something, they also positioned themselves as agreeing or disagreeing with the stands of others. Yet the triangle is also apt for describing coordination regarding perception and knowledge, or what linguistic anthropologists call "epistemic stance." As a presentation of evidence, the Collector's tour of the pollution was an effort to coordinate both epistemic and ethical stance at once. When Manamur protestors directed her attention to the gunk in the water, the foam on the riverbank, or the stench in the air, they were also inviting her to share their evaluation of these sensory objects. They sought to draw her into a shared experience of sights and smells and, thereby, into a shared encounter with injustice. Pointing with both hands and words, and embedding what she saw in stories of stinky wells, late night coughing, skin rashes, and cancers, they led her in tracking the movement of the pollution through their lives.

### *Directness Ritual*

During the Collector's visit, it was difficult to tell how much she was persuaded by Action Council members' tour of pollution. Her interaction with local activists seemed cautiously quiet and diplomatic, and her statement to the TV reporters was ambiguous. Yet my research assistant Ahmed and I knew that this tour could be powerful because we had both been through it ourselves.[12] In this regard, Ahmed's first visit to Manamur had been remarkable. He went there shortly after the police baton charge (see chapter 2) and, because my friends had warned me that any foreigner visiting would likely be deported, he decided to go alone. Before leaving, he expressed to me that he intended to take the role of an objective researcher, withholding judgment about the pollution. But when he returned, all of that had changed. He told me he was certain that the protesters' claims were true, and he felt that he should join the campaign. How, I asked, had this change happened?

"They all told me 'there is so much pollution in the water,' and I heard them," Ahmed explained. "But when I saw it directly, then it was really true. Because in the water there was this smell. A smell was there, and there was the color change of that. So it must've been when I saw all of that."

The term Ahmed used to describe the power of his encounter with the pollution in Manamur, "directly" *(nēriṭṭ),* was commonly used by Action Council members as well. An adverb derived from the Malayalam for "straight," *nēriṭṭ* is also commonly used to distinguish face-to-face encounters from, for example, phone calls or letters. Local activists often used this term to describe their own experiences with pollution, particularly the experiences that had led them to join the campaign (see chapter 2). Likewise, they talked about visitors seeing the situation in Manamur "directly" during guided tours of the pollution.

This idiom of directness belies the work of intersubjective coordination that was so apparent in Action Council members' displays of pollution. Observing such tours every day, Ahmed and I became aware of how much work protestors put into carefully orchestrating and facilitating these "direct" encounters. Several times a day, activists led students, religious leaders, and government officials to the same spots; made the same arguments about acid, asthma, and cancer rates; and told the same story of a wedding reception that had been canceled due to the bad smell. But calling such displays "direct" suggested an unmediated encounter between the visitor and the pollution, an encounter whose persuasive power hinged on the properties of objects, rather than relations between subjects.

Anthropologists studying similar claims of direct access to reality—what Mazzarella calls the "politics of immediation"—have observed that such claims are always contradictory insofar as they deny the social processes by which "directness" is produced.[13] But the Manamur protestors' claims about direct encounters with pollution are not best heard as denials that pointing, commenting, narrating, and other acts of mediation were involved; rather, they are better understood as descriptions of what all of this mediating effort was about. When Ahmed told me about his first encounter with pollution in Manamur, he described how the locals had prepared him and guided him. But what really convinced him to join their cause, he claimed, was what they guided him to. The smell of it. The color. He did not experience the pollution this way because he had joined the campaign. Rather, in his self-understanding, it was his encounter with pollution that made him want to join. This is how the pollution tour was supposed to work.

The physical properties of the gelatin factory's pollution were important to directness, but they were not enough. In Chapter 2, I described how fighting the gelatin factory amplified local protestors' everyday experience of "the smell," creating a feedback loop that bolstered their commitment to the cause. Similarly, in the directness ritual, the local protestors worked to immerse their guests in a dense sensory experience of pollution. To do so, they drew on many sensory cues: a scent on the air, a yellowish foam at the edge of the rice paddy, a slipperiness when well water was rubbed between thumb and forefinger. Yet these properties were not always available. The smell could vary with the level of production in the factory, the direction of the wind, or the sensitivity of noses. The foam caused by factory effluent was not necessarily, to an untrained eye, distinguishable from foam stirred up by tilling the soil. Other sensations were even more difficult to access: the protestors could lead the Collector down to the edge of the river paddy or the irrigation canal, but they could not make her wade in. She would not feel the itchiness. This is why it was so important for Peters to stand in the water all morning—his dizziness and nausea could render the potency of the pollution visible to others. And the most severe bodily experiences—the chronic congestion and coughing, the cancer—were also those most difficult to convey to visitors. They entered the directness ritual only as stories, anchored only by deictics (the child in *that* house, *there*) to a visitor's sensory surround.

There is a clear parallel between local Manamur activists' emphasis on directness and the radical environmentalists' efforts to put principles over relationships, described in chapter 3. Claims about directness similarly minimize the role of social influence on one's moral views. In the stance triangle's terms, they stress the evaluative side of ethics over the intersubjective, relations to objects over relations to other subjects. The two ethical projects are also similar in their limits. In chapter 3, I detailed how the intersubjective side of ethics continually reasserted itself into radical environmentalists' efforts to make change. Here, it is clear that direct encounters with pollution are thick with social influence.

Yet the differences between these two modes of object-oriented ethics are important. Radical environmentalists sought to expand moral concern beyond particular social groups and, likewise, to cultivate a moral subjectivity free from group influence. Local activists did not share this expansive aim, nor were they concerned with forming moral subjects. Rather, pollution tours sought to hold the intersubjective side of ethics in abeyance during specific moments, or in specific social scenes, so that the objects of evaluation could speak for themselves. Through the smells and colors, visitors would briefly inhabit the Action Council's world, and this experience would convince them to join the fight.

To understand when and why claims of directness were so important to Manamur protestors, consider another pollution tour, later on the day of the Collector's visit, when the ritual of direct encounter plainly failed. Following up on the Collector's call for a "technical report," a researcher from the state government's Pollution Control Board (KPCB), came to take samples of the murky water in the paddy. Initially, Action Council members were skeptical or even hostile—earlier government studies had not all been in their favor. Some protestors shouted at her in rough language, and she later claimed they threatened her. Nonetheless, when the researcher requested that Action Council members show her where to take samples, a few protestors astutely perceived this as an opportunity for a direct encounter. They took her to the same spots as the Collector, pointed to the same pathways of flow, and told the same stories of cancer and coughing. Others soon joined them, so that a small crowd gathered around her. But when they appealed to her sympathy, attempting to embed what she saw, smelled, and sampled in the broader context of their lives, she only bent more closely over her notebook, shielding herself from their attempts to catch her gaze, and said nothing in response to their stories of suffering. Her lack of response seemed to agitate the protestors, who became more and more insistent in their appeals. As she was getting into her car, one of them bluntly accused her of supporting the gelatin factory.

"That's what you say, isn't it?" she replied flatly, chuckling as she climbed into the car.

With this question, "That's what you say, isn't it?" the government researcher deflated the persuasive force of the directness ritual. In one sense, her reaction paralleled the rhetoric of directness; she implied that she was only concerned with taking samples of the sludge, not with the protestor's words—i.e., she attended to the objects of evaluation, not the evaluative positions of other subjects. Yet by drawing attention to the mediating role of campaign participants' speech, she also undermined their attempts to package the sensory experience of the sludge into their own stories of injustice.[14] What they said would have to remain only what they said, disconnected from the sights and smells that had swayed so many other visitors. She had made use of their guidance for the purpose of taking her samples, but she would not be drawn into their experiences of suffering. Instead, she sought to access moral objects by other means; her study would supply her evaluation.

## *People's Science*

Action Council members' failed attempts to persuade the District Collector and the government researcher were symptomatic of a more general limit on the persuasive force of sensory displays: they depended on ideologies of immediacy that, though potent in face-to-face interaction, were not operative in state agencies. But pollution tours were not the only way in which Manamur activists attempted to mobilize evidence. From the very beginning of the campaign, the Action Council made numerous efforts—often in collaboration with sympathetic academics and NGOs—to back up its claims with scientific data on lead levels, air pollutants, cancer rates, and other measurable indicators of health hazards. Action Council leaders saw scientific research as especially important to their legal challenges to the factory, a strategy they pursued alongside street protest.[15] Yet they were skeptical of research commissioned by government agencies or the courts. By the time the District Collector arrived, numerous such studies had been conducted. One had recently found elevated levels of some heavy metals, but the government commission had ultimately determined the data was inconclusive and called for further study. When the Collector called for a "technical report," they only heard this as one more attempt to stall government action and stymie the campaign's momentum. But the problem was not science as such. The problem, the protestors said, was how to mobilize science that aligned with their experiences of the pollution—how to make a science of the people.

Environmental movements commonly struggle with contradictions between people's experiences of injustice and the findings of scientists.[16] Such contradictions are present in other social movements as well, but environmental movements are unusual in their degree of entanglement with evidentiary practices from the physical sciences.[17] As Sylvia Tesh shows in a study of US environmental justice movements, there is no inherent incompatibility between lay ways of knowing and the forms of evidence used by experts—the experiences of so-called lay protestors are, often, already informed by scientific research.[18] This was true in Manamur: when Action Council members complained of coughs and cancers, they did so against the background of existing studies of toxins in the air and water. Nonetheless, it is often difficult to harmonize these two modes of moral appeal.[19] On the one hand, Action Council members' claims to be "the people" were rooted in narratives of suffering and rage. On the other hand, activists' uses of science demanded a process that could detach molecules and micro-organisms from these narratives as well as a textual form that silenced tones of rage.

Opposition between the people and science has not always been regarded as inevitable in Kerala. From the early twentieth century, Kerala's Communist movement sought to ground its politics in absolute commitments to both the masses (*bahujanaṅṅaḷ*) and to science. In the

1960s, this Communist ethic produced the Kerala people's science movement (*Kēraḷa Śāstra Sāhitya Pariṣatt*, or KSSP), which sought to spread literacy, educate the masses in elementary scientific theories, and harness science in service of the people.[20] The people's science movement was by no means epistemologically pluralist; scientific findings were taught to the people, not questioned in light of popular beliefs. But while science provided the means of change, the aim of people's science was to defend the interests of the people against capitalism, corrupt governance, and other forms of injustice. Throughout the late twentieth century, the KSSP popularized this idea by conducting scientific demonstrations, seminars, and children's programs in villages like Manamur, but also by taking a lead role in some environmental conservation movements.[21] By the time the Manamur Action Council formed in 2008, the possibility of people's science as a tool for environmental justice was very much on the table.

Nonetheless, efforts to produce people's science in Manamur consistently fell short in one of two ways. On the one hand, the campaign and its supporters struggled to produce science that was recognized as authoritative by state agencies, the courts, or even the campaign's outside supporters. On the other hand, the Action Council had difficulty producing scientific results that affirmed their members' experiences of suffering. A true people's science would, ideally, be able to meet both of these challenges at once.

The prime example of the first challenge was a study by an NGO called *Jananīti* ("People's Justice"). Founded by a former Catholic priest-turned-lawyer, *Jananīti* offers free legal aid and puts out a small-circulation magazine on human rights issues that overlaps in content and vision with *Kēraḷīyam*.[22] But the NGO specialized in producing "investigative studies" on issues related to people's protests. Action Council members often said that further research was not necessary because there was already an abundance of scientific evidence on their side, and the *Jananīti* study was the primary basis for this argument. The *Jananīti* study confirmed all of the health claims recited in the pollution tours. Because it was published in the first years of the Action Council, it is likely that its findings influenced the framing of these claims. In particular, the study found high levels of heavy metals in the water, and campaign participants often described these metals as the reason for high levels of cancer and other illnesses in the village.[23] Thus, this study had been adopted by the campaign, as it were; its claims blended harmoniously into the voice of the people.

Unfortunately, beyond its influence on the views of Action Council members, the *Jananīti* study had little impact on the trajectory of the campaign. The reasons for this failure were a topic of some debate. When it was initially released, the company held a press conference declaring that the report was obviously biased. Of course, such criticism was only to be expected. Yet the study had many critics among the Action Council's members and supporters as well. Critics aligned with the campaign described the study's problem as an excess of passionate emotion (*vikāram*). The study, they argued, was too closely and obviously aligned with the perspectives and emotions of campaign participants; it read like an activist report more than a scientific report.

James, the former Catholic priest who heads *Jananīti*, acknowledged that the report was emotionally charged, but he did not see this as a shortcoming. "It is an activist report," he explained, "but, facts are there. You can't deny it. See, that means there is a kind of emotional involvement in that. It is not a false report. All the same, it is not a neutral [report]. It feels for the people." James presented emotional involvement as a distinct aspect of *Jananīti*'s report that should not be understood as tainting the presentation of facts. Indeed, he said that from the start he had publicly confirmed the company's contention that the report was biased. He openly took sides. But he had also challenged the company to prove that the report's claims were false. He depicted the study's moral alignment with the Action Council as immaterial to the authority of its findings. The voice of the people might sing through in his report, harmonizing with the facts presented there, but the facts still spoke with their own voice.

Critics of the report argued, however, that *Jananīti*'s emotion rendered its facts unconvincing. It was not necessarily the case that they believed the facts were tainted by bias, as the company claimed. What they argued, rather, was that when it came to persuading non-activists to support the campaign, perceptions of bias were as much a problem as bias itself.

Such was the criticism levied by Fahad, a young professor at a nearby engineering college and a member of Solidarity [Eng], a Muslim youth group that focused its activism on supporting environmental justice protests throughout Kerala.[24] During much of my research, Fahad was working on a new study of the Manamur pollution that was meant to make up for the inadequacies of the *Jananīti* study. Its failing, he argued, was in the feelings it aroused in its presentation of that evidence: "From the beginning itself, anyone reading that report will feel that it is a report meant to help the protest win. But the report we are preparing, without feeling that it is the protest's report—eh, not without feeling, not like that . . . This [is] evidence. Facts."

Fahad's version of people's science ran into problems of another kind. Having never conducted a study of pollution before, Fahad researched the procedures for environmental impact assessments set out by the US Environmental Protection Agency. He adopted many of their research methods, such as mapping out zones for sampling and taking samples in triplicate. In addition, he structured the report with what he called "the pattern of science"— an abstract, introduction, literature review, methodology, results, discussion, abbreviations list, and references—all in what he called the "standard" order. Late in my research, I visited him at the college, and we paged through the completed report together: two thick, colorful volumes of tables and charts, sandwiched between glossy white plastic with brass clasps at the corners. The report was meant to persuade scientists and to be usable in court cases, he explained. Although he anticipated that I, as a fellow scientist, would appreciate its findings, he acknowledged that it would not be readable by common people, *sādhāraṇakkār*. Its scientific form, which made it so powerful in other respects, also made it difficult to communicate to those it was meant to serve. For that reason, Fahad planned to make a PowerPoint summarizing the results, which he would present to the Action Council and its supporters before publication.

But Fahad's PowerPoint did not go as planned. A week after our interview, he stood before dozens of participants in the Manamur campaign in a large hall in the nearest major town, defending his findings. For most of the presentation, the audience was fairly quiet. Local campaign participants nodded and murmured as Fahad presented maps of the sampling sites. One attendee, a scientist with the fisheries department, raised some issues with some of his conclusions about microorganisms in the samples, but Fahad's answers seemed to satisfy him. Near the end, however, there were more questions along these lines, and a problem began to take shape. Fahad had found levels of certain bacteria that could be hazardous, but he had said little or nothing about heavy metals or cancer. He had found pollution, but not the pollution that the campaign had been protesting all this time.

Fahad countered objections by suggesting that his results opened up new possible directions for the campaign, directions that would be grounded in firmer evidence. But his audience did not seem convinced. Action Council members, in particular, were concerned that Fahad's study would invalidate their longstanding claims. The presentation ended on an ambiguous note. The next day, I received news that the release of Fahad's study had been postponed indefinitely. When I departed Kerala several months later, it still had not been released.

Fahad, like James, believed that people's science was possible because of the obdurate power of facts: the impossibility of explaining them entirely in terms of empathy or ethical alignment. James believed that the facts in his report carried their own persuasive force, irrespective of the sentiments it conveyed or the allegiances it proclaimed. Fahad attributed the power of facts to the processes by which they were produced and the forms in which they were presented. Yet both shared the belief that science had the power to translate Manamur locals' experiences of suffering into words that could not be written off as—to recall the words of the Pollution Control Board researcher—only "what you say." Like the pollution tours, both versions of people's science were techniques to divest the evaluative side of ethics of its intersubjective baggage. Just as the colors and smells had persuaded Ahmed, so both studies sought to make the pollution speak for itself.

James and Fahad were not alone in turning science to such purposes. Though they differed from one another in their ideas about facts, both their efforts drew upon widely circulating philosophies of scientific objectivity. In the philosopher Bruno Latour's terms, both efforts used techniques to make objects *object*—that is, to speak up and denounce the injustice in Manamur.[25] Traces of these philosophies can also be heard in the District Collector's statement to the TV reporter. Her interaction with protestors enabled her to know the pollution "in a layman's way," drawing her into the intersubjective coordination that the Action Council sought. But this alignment did not extend to the government, which can only know the pollution through "technical reports" of the sort that the researcher's sample-taking, later that afternoon, was intended to produce. Through people's science, the Action Council sought to traverse this epistemological boundary, carrying their evidence of pollution beyond the reach of their own voices, beyond "the smell."

Yet scientific objectivity was difficult to bring into sync with local activists' experiences and claims. People's science in Manamur either failed to effectively assert itself as a reliable channel for objective reality or, alternatively, threatened to clash with the voice of the people. The pollution tours did not have this problem, but they could only reach so far. They might sway a small number of powerful visitors, but this might still be difficult to translate into government intervention. Thus, producing evidence of pollution, whether sensory or scientific, was severely limited as a tactic of persuasion. To make their reality compelling beyond Manamur, the Action Council sought to mediate directness in other ways, beyond the limits of the senses and the constrictions of state-dominated discourses of the scientific. In doing so, they also addressed more distant audiences, beyond the District Collector or the Pollution Control Board. They turned to the public for support.

## STAGING SPONTANEITY

The main road of Manamur was packed with people, protest slogans shook the air, and TV crews were capturing it all. The moment Action Council members had been awaiting for days had arrived. As the hunched, white-clad figure of one of Kerala's most famous politicians came walking around the bend, Vijayan and other Action Council leaders rushed to meet him, taking his hands in warm greeting and gently, but firmly, steering him toward their protest tent (*samarapantal*) beside the factory gate. Other Action Council members cheered from the tent's concrete platform, waving the cardboard signs they had made that morning. But even as he approached, those men walking behind him, some of them likewise draped to their ankles in white, stepped quickly forward, barred his way, and raised their voices above the hubbub. They pointed him down the road to a different tent, a huge red tent constructed a few days before by his own party. For a moment, the silver-haired man slowed to a stop, squinting around in the noon sun, as if the opposing forces of all the steering hands and coaxing voices had canceled one another out. But his face had already turned toward the other tent, the red tent, and now his fellow party members closed around him, herding him along, while the Action Council members dropped back in silence.

. . .

In Kerala, to protest is to "go out into the street" (*teruvil iṟaṅṅuka*), and the campaign against the Manamur gelatin factory was largely conducted in streets—whether that meant marching from town to town, holding sit-ins in front of the factory gate, battling with police, or crowding on a bridge above a polluted canal. As performance spaces, Kerala's streets offered audiences near and far, in-person and virtual. Early-twentieth-century anti-caste movements had marched to open existing roads to all castes,[26] preparing them as conduits of public life—dense channels of sights and sounds through which bodies of all sorts pass: friends and enemies; kin and strangers; environmentalists, politicians, and anthropologists—not to mention carefully arranged rows of dead fish.[27] But displays in the street—whether in the tent, at the factory gate, or on the bridge over the canal—were never only meant for passersby. A speech, a march, or—especially—a fight would not only draw the gaze of bystanders but of their phones, and soon videos would be making their way, via Facebook, WhatsApp, or YouTube, to more distant eyes and ears. A crowd in the street could also draw TV and newspaper reporters, who might interview activists and take footage of dead fish, foamy riverbanks, or welts from a police officer's baton.

As with other street protests in Kerala, the hub of this activity was the protest tent, a raised concrete platform reaching right out over the asphalt, backed by the factory wall and roofed with plastic sheeting. As described in this book's introduction, the tent was a gathering place—a place to hold meetings and strategize, but also a place to celebrate and chat with friends. This was not the same, however, as meeting at a friend's home, going to a church festival, or even attending a function on a school ground. The tent was built like a stage, with an open fourth wall facing into the street, and when they held marches or welcomed distinguished visitors, the Action Council used it like a stage.

When a famous politician or civil society figure came, a podium was set up on the platform with two rows of chairs behind it. Chairs for the audience were set up in in front of the tent and on the far side of the road, and often a crowd would grow around these chairs, blocking traffic entirely. Most such programs would begin with a demonstration in front of the factory, where slogans and demands for justice could be addressed directly to the closed gates. The factory's managers never came to the gate to hear these demands, and no one expected they would; the audience that mattered was much farther away. The floor of the Manamur tent was often littered with newspapers and magazines carrying the latest reporting on the protests. Sitting on the edge of the concrete platform, protestors passed the time by checking the latest posts about the campaign on Facebook and uploading their own.[28]

People's protests did not have a monopoly on street politics; the Manamur Action Council's tent was not even the only protest tent on their road. During one of the most intense periods of the campaign against the factory, other tents began popping up along the main road in Manamur. It began when the effluent pipe from the gelatin factory broke, spilling putrid black gunk into an adjacent rice paddy—the same spill that brought the District Collector to the village. First, the Action

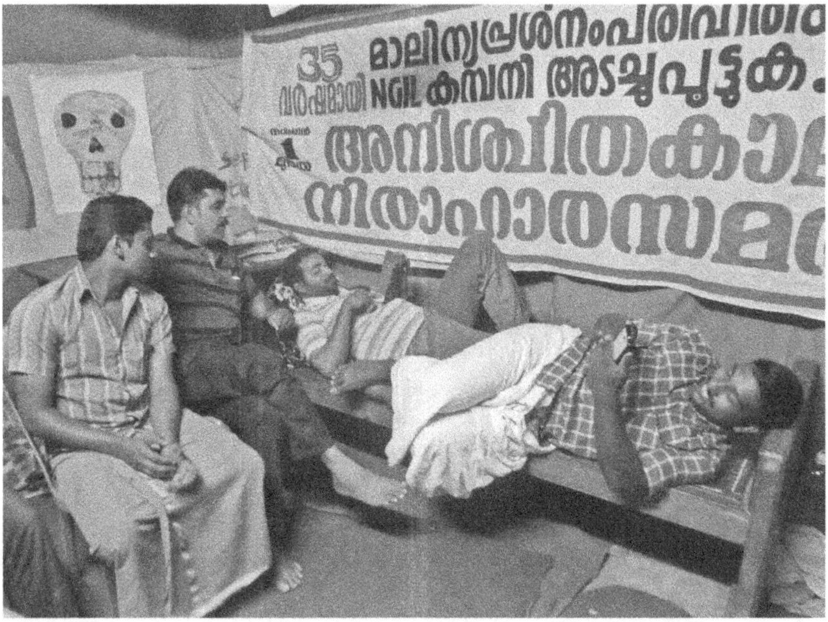


FIGURE 17. On a quiet afternoon, young male protestors check the latest news and social media about the protest.

Council itself built another, similar pavilion, with a wooden platform instead of concrete, along that stretch of road as well, hoping to take full advantage of the new sights and smells. But this was immediately joined by another tent, just a hundred meters away, set up by factory workers protesting in favor of the factory. A few days later, the CPM constructed yet another tent—and thereby launched their own anti-pollution campaign, conspicuously *not* aligned with the Manamur Action Council—at a junction halfway between the factory gate and the paddy field. What this last tent lacked in location, it made up for with noise and spectacle. Their tent's platform was about three times as large as that in either of the Action Council's tents, and bright Communist red. They wired loudspeakers down the road in either direction, flooding the Action Council tents with their own speeches and songs. For a time, at least, the CPM had stolen the stage.

The inability of the Action Council to compete with the street protests of Kerala's major political parties was never more apparent than on the day the famous politician—V.S. Achuthanandan, a revered leader of the CPM with the rare distinction of being respected by many outside his party as well—turned and continued down the road. Yet

FIGURE 18. CPM members gather in a newly constructed tent down the road from the Action Council tent.

even on that day, Action Council members assured me that they were not really concerned about being sidelined by the CPM protest. The other tent might be louder and more spectacular, but that would not necessarily make it a more powerful performance. After all, the power of the Action Council's roadside displays did not lie in their volume, but in their logic of moral appeal.

The Action Council's success in enacting the people had little to do with how many people they could assemble. Rather, being the people was primarily a matter of distinguishing their campaign from the broader ecology of protest, which was dominated by the protests of political parties, trade unions, and religious groups. These groups assembled in the streets in order to demonstrate their strength as mass actors representing large portions of the populace. The Manamur Action Council, like other environmental justice movements in Kerala, did not have the numbers to back such claims. Yet the discourse of "the people" provided a means of turning this seeming weakness into a strength. Rather than gathering as many bodies as possible, they sought to differentiate their protests from the politics of parties and partisans by manifesting the *whole* people, a task for which no mere portion of the people, no matter how large or loud, could suffice.[29]

This rhetorical strategy centered on the comportment of bodies. Through displays of sacrifice and uncontrollable rage, the Manamur protestors foregrounded their bodies' vulnerabilities and incapacities, marking their protests as pure reactions to the harm done to them by others—as eruptions of the people, desperate for justice. As with their uses of evidence, these appeals hinged on erasing protestors' own values, powers, and agendas. Rather than the powerful collective subjects manifested by the CPM and other organizations, they sought to make themselves a sympathetic moral object for evaluation by the public.

### *Displaying Dominance*

To understand performances of vulnerability in people's protests, it is necessary to first understand the broader culture of protest (*samaram*) within and against which some protests are distinguished as protests of the people (*janakīya samaraṅṅaḷ*). Protest, or *samaram,* is a pervasive and unavoidable part of everyday life in Kerala. On any trip of more than a few kilometers, you can expect to meet a *samaram* or two along the way. From the window of a bus, you will see marchers hoisting their flags and chanting slogans. Occasionally, but more often than most Malayalis would like, you will not be able to set out at all because a political party or major trade union has called for *hartal*—a total shutdown of shops, auto-rickshaws, buses, or even the roads themselves. While nearly everyone complains about street protests, they nonetheless go on (see introduction).

Still, even if many Malayalis at times take to the streets, some do so more frequently and effectively than others. In this regard, the CPM reigns supreme, holding larger and more frequent demonstrations than other parties. The CPM is particularly adept in its use of *hartal.* While bus or taxi unions occasionally call *hartals* focused specifically on transportation—such as prohibiting the use of vehicles—full *hartals,* usually called by political parties, not only block travel through roads but also forbid the opening of storefront shutters that face the road. During the *hartal,* young male party members roam the streets, sometimes with sticks, threatening to beat those who transgress the prohibition. Thus, this form of political action requires a large, dedicated membership. But a successful *hartal* elicits collaboration, if not support, from actors outside the party as well. Newspapers, TV channels, and radio stations must inform the populace of the *hartal's* region, duration, and specific restrictions.[30] When a major party calls a *hartal,* police will walk the streets with their own sticks, enforcing the party's prohibitions in the name of "keeping the peace." In this way, the *hartal* becomes "official," even to the point of being treated by much of the populace as a kind of state holiday.

When people's protests take to the streets, they draw from the same social movement repertoire as the CPM, including *hartal.* But their *hartals* rarely have the same effects. In the aftermath of the police baton charge in Manamur, for example, the Action Council announced a *hartal* for the local district of Thrissur. That evening, the woman who had been helping care for my infant daughter called to say she would not be able to come. But Adarsh informed her that she should not worry; since the *hartal* had only been called by the Action Council, it was unlikely any of the buses would stop running. And indeed, the next day, there was no sign of any change in traffic, let alone closed shops. Perhaps some shops closed in Manamur village itself, but villagers told me later that there had been very little difference there either. Many people in the area had not even heard that a *hartal* had been called. Three weeks later, the CPM held a demonstration in Kerala's capital city against alleged corruption by the ruling Congress party, and the contrast was palpable. They swarmed the streets around the Secretariat, the center of state government administration, effectively shutting down not only commerce but the government itself.

What most obviously differentiates CPM protests from people's protests is numbers: the CPM had the power to bring in enough bodies to patrol the streets, if not to clog them entirely. Most people's protests simply do not have this capacity. While they aspire to gain sympathy and support from non-local actors, they can rarely rally the numbers needed to take over the streets. When they call *hartals,* they cannot enforce them. They march, but they cannot swarm.[31]

A party's ability to, by turns, empty roads (in *hartal*) and clog them (in marches and rallies) was more than simply a show of membership; it was a show of dominance over public life.[32] In Kerala today, *hartals* are rarely, if ever, called by the ruling party; they are called by the party that is in the opposition. By taking over public roads, they challenge the ruling party's claim to represent the populace.[33] For a time, the opposition party asserts its legitimacy as the real ruling power.

FIGURE 19. Local Action Council members conduct a torchlit march through the streets of Manamur.

Thus, though it may be a matter of numbers, the power of CPM street protest is not merely a matter of brute force; it is a symbolic power that depends on extensive mass mediation. Taking over the streets is also about taking over the TVs. And this, too, differentiates people's protests from the protests of the CPM and other parties. Both sought the attention of mass media, but party protests could dominate the airwaves to a degree not possible for people's protests. The police baton charge brought more media attention to Manamur than at any time before or since. Yet even on the day of the violence, the campaign against the gelatin factory was not the top story on Kerala's many twenty-four-hour news channels. On the contrary, on that day, the news channels were already incessantly covering the corruption scandal that would culminate in the CPM "siege" of the Secretariat. It would be three weeks before the CPM took over the streets in the state's capital, but the airwaves were already theirs.

*Displaying Sacrifice*

Given the CPM's advantage in numbers, I was initially surprised that the Action Council members so easily shrugged off the red tent down the road. At first, I thought they might simply be content to see pressure building against the factory, whether it sidelined their own campaign or not. But this was not their reasoning. On the contrary, what reassured Action Council members, they told me, was the certainty that the CPM campaign would not last long.

The notion that the CPM would, before long, be paid by the gelatin company to stop protesting was consistent with Action Council members' own narrative about their campaign. Between the opening of the factory in 1979 and the formation of the Action Council in 2008, they said, many other campaigns had come and gone. These other campaigns had not lasted because their leaders had eventually been bribed by the company; sometimes it was said that they had only protested in order to attain such bribes. The Action Council participants differed from these earlier efforts in that they had already lasted much longer. Despite great hardship, they had endured. And their endurance was a sign of the sincerity and commitment with which they pursued their cause. It marked the difference between the vying claims of parties to represent the populace and a true uprising of the people.

This semiotics of endurance was part of a broader performative strategy that centered on displays of bodily danger or harm. In contrast to what some called the "power politics [Eng]" of parties, activists in people's protests displayed their corporeal vulnerability. In Manamur, displays of physical risk and suffering included, of course, the tours of pollution in which activists told visitors about skin rashes, late-night coughs, and cancer. It also included demonstrating in front of the factory gates until they were beaten by police. More mundanely, simply occupying a tent for long hours was treated as a public display of sacrifice. Activists' willingness to endure lost wages, lost friends, and the pure exhaustion of years of campaigning could be a sign of their commitment to the cause. Thus, protestors in Manamur considered it crucial to keep the protest tent occupied at all times. If the tent was empty, the people were not protesting.

Hunger strikes were a central tactic in the broader repertoire of such displays of physical risk and suffering. During the period of my fieldwork, protestors in Manamur conducted multiple hunger strikes, each with different tactics. In the ramping up of protest activity prior to the

July 2013 police baton charge, two prominent women in the campaign, including one member of the local *pañcāyatt*, undertook an indefinite fast that culminated in their arrest by police. In the aftermath of the police violence, one Action Council leader attempted to conduct further hunger strikes, but few participated—with so much media attention already on the campaign, and public sympathy seemingly very high, the general feeling was that hunger strikes would be an unnecessary burden. Several months later, however, after media attention had waned and solidarity efforts by environmental activists from outside of Manamur had ceased (see chapter 2), the Action Council again conducted a hunger strike—this time organized as a relay—in an effort to recapture public sympathy.

Such displays of suffering are traceable to the mode of nonviolent political action Gandhi called *satyagraha*—a term Manamur activists self-consciously use to describe their campaign. Variously glossed as "truth force," "truth struggle," "passive resistance," or "nonviolent resistance," *satyagraha* is not so much a specific practice as a political philosophy, and its meaning has been much debated by scholars.[34] In Manamur, the meaning of *satyagraha* is far less abstract; people use it to reference the continual occupation of the tent, sometimes while fasting. Nonetheless, hunger strikes in Manamur, like Gandhi's own fasts, tap into a semiotics of sacrifice with deep roots in India. They draw on religious traditions of fasting as a form of self-renunciation—welcoming hunger of the body as a means of transcending the desires of the self.[35] Yet this is not self-renunciation in a religious mode, aimed at spiritual liberation. Hunger strikes in Manamur were not the culmination of a good life, but the mark of an unbearable life.

Not all bodies in Manamur were seen as equally capable of eliciting sympathy. As others have noted, in India bodies are commonly understood to be fundamentally unequal: bodies and bodily substances are the ground, both metaphorically and materially, for the maintenance of social hierarchy.[36] Caste hierarchy figured prominently in precolonial hunger strikes, in which Nambudiri brahmins would fast to intimidate others into acceding to their demands.[37] The Manamur relay hunger strike differed from this tradition in that it involved the sacrifice of diverse bodies—across class, caste, and gender. But it also reflected the differential value of these bodies.

The relay hunger strike was originally supposed to begin on a Monday morning, but when Ahmed and I arrived at the Manamur protest tent, there was no one there. It was not long before a few of the usual men began to arrive, but the mood was glum. They said that the relay hunger strike had been postponed because Stevenson, who was supposed to take the first leg of the relay, had been advised by his doctor not to participate. A local celebrity with political aspirations, Stevenson had warned the other activists about his high blood pressure and diabetes when he accepted the inaugural role. At the time, everyone had laughed, joking that they all had high blood pressure and diabetes. No one was joking anymore.

Local activists wanted Stevenson to inaugurate the rally because of his renown, and when he could not, they searched for some other such public figure. Moreover, later on, some activists complained that the hunger strike was failing because they had not been able to recruit famous participants. It was pointless to lead a hunger strike with only local people, they said, because no one would pay attention.

While a more famous or higher-status person's body might draw more attention, the semiotics of sacrifice was available to less esteemed bodies in Manamur as well.[38] Unequal bodies could, to some extent, substitute for one another in the relay hunger strike because the performance of self-renunciation depended not on the social value of one's body but on the perceived worth of one's body to oneself. Willingness to sacrifice one's body for a cause was powerful because of a perceived incompatibility with the interestedness, calculation, and artifice of everyday politics. As Bargu notes in a study of hunger strikes among Turkish prisoners, the body's "deployment only by way of its destruction" seemingly "obliterates instrumental rationality."[39] One who puts one's body at stake, the notion goes, cannot simply be posturing for political or financial gain; who would be left to gain? Sacrifice of the body becomes the ultimate sign of sincerity.

Even as the Manamur Action Council used displays of sacrifice to differentiate their campaign from political parties and their displays of dominance, their efforts were always under threat from the political party's own use of the Gandhian tactical repertoire. Political parties did not limit their protests to *hartals*. In the initial days after the CPM raised its red tent in Manamur, they held their own hunger strike featuring prominent leaders from the party. This threatened the Manamur activists' efforts to distinguish their campaign as a protest of the people. "*Satyagraha,* hunger strikes, *hartals,* and road blocks are all tools of protest," explained one activist. "But when the political parties use them it weakens them for everyone. Their sharpness is lost. For Gandhi, the hunger strike was a complete offering of oneself."

Both the CPM and the Manamur Action Council used hunger strikes to tie their protests to the original, archetypical people's protest. But unlike in the politics of marches and *hartals,* in this semiotic economy, the Manamur protestors' lack of rank-and-file multitudes, vast financial resources, and high-level contacts gave them the upper hand. Here, being the people hinged on performing vulnerability—on being perceived as making "a complete offering of oneself." To be the "whole people," the philosopher Rancière argues, a set of actors must be defined not by their particularity, but only "in the name of the wrong done them by the other parties."[40] *Satyagraha* marked the Manamur tent as different from the huge, red tent of the CPM not by asserting the strength of the Action Council as a separate organization, with its own flags and anthems, but by showcasing the harm done to Action Council activists by others.

By sacrificing their bodies, these activists sought to de-emphasize their own organizational affiliations and political aspirations, making their campaign appear a natural outgrowth of the injustice committed by the gelatin factory. Before he joined the campaign, Stevenson had been known as a loyal member of the Congress Party, but after joining he gave up his white shirt and *muṇḍ* (the traditional attire of politicians of his party), seeking instead to present himself as another victim of pollution, another body on the line. In other words, even as the power of his fast was tied to his social status, it also depended on distancing himself from his image as a politician.

Eventually Stevenson did, with his doctor's assent, launch the relay hunger strike by being the first to fast. The strike began with an inaugural ceremony, well attended by the press, in which speakers declared the desperation of the people in Manamur and their willingness to fast until death. When, after one week, Stevenson did pass the torch to the next faster, he was hauled off by the police. This was by no means a sting operation. Campaign leaders informed the police that Stevenson was gravely ill and insisted on fasting unto death, and when the police arrived, Action Council members helped to load him onto a stretcher and carry him to the police car. Media was informed ahead of time. Here, the performance was clearly not aimed at the government, even as the arrest presupposed the state's commitment to asserting its power to "ensure, sustain, and multiply life."[41] The addressee was the public, by way of the media, by way of the police. The arrest ratified Stevenson's fast as a "complete offering" of his body rather than simply a tactical ploy.

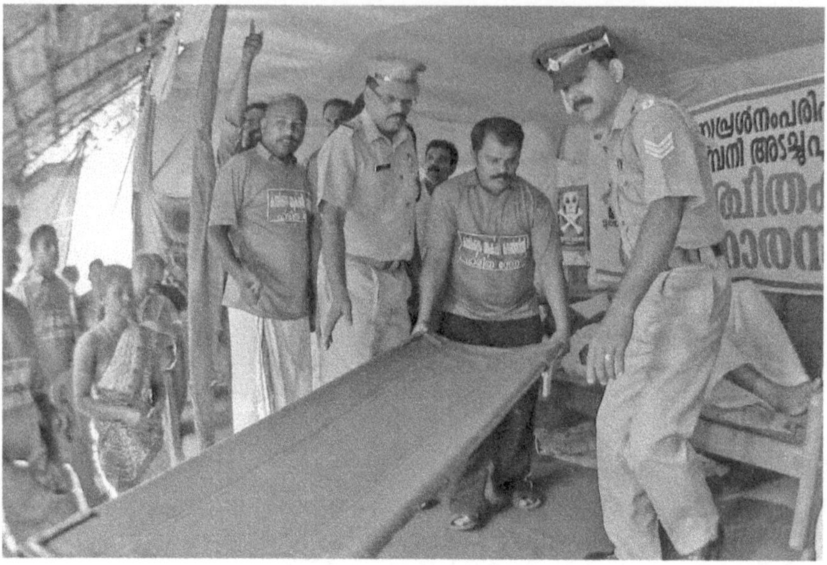

FIGURE 20. After six days of fasting, Stevenson is arrested by the police at the protest tent.

## *Displaying Rage*

Despite the centrality of *satyagraha* to their campaign, the Action Council also engaged in certain acts of violence. Periodically, some campaign participants, usually men, would do things like get into fistfights with factory workers or throw rocks at police. Campaign leaders did not openly describe these as the Action Council's activities, nor did they plan for them in the same way they planned for speeches, marches, or for ensuring the tent was occupied. This is not to say that violence was unexpected—there was frequent boasting among men in the tent about what they "would do" or "would like to do"—but the actual episodes of violence were generally described as spontaneous and uncontrollable outbursts. The quarrel with the reporter on the bridge was one such outburst, or nearly so. In the end, the reporter left without any blows thrown, but everyone said it had been a close call.

"We cannot help ourselves," explained Satish, who had been foremost in the fray. "Rage (*vikāram*) is the strength of our campaign. But, because we have experienced so much suffering, it is so strong that it makes us do things like this."

*Vikāram,* the term I gloss here as "rage," can be used to describe a broad range of emotions—such as anger, sadness, or lust—that have an overwhelming effect. *Vikāram* is passionate emotion: it can drive a person to act in ways that they would not when level-headed.[42] Local activists often described their *vikāram* with pride, calling it the source of the campaign's strength and tenacity. They also used the term to describe themselves at times when they seemed to anticipate that I might object to their actions, for example when they became openly hostile with factory workers or journalists. Likewise, when some of the Action Council members used threatening language with the government researcher who came to test the water, one of them told her that she should not hold it against them. "It is only because of their *vikāram*," he said, "that the people say such things." For the Manamur campaign participants, *vikāram* was induced by prolonged suffering. It was an anger that motivated them to fight, but it could also make them lose control.

Among supporters of the Action Council, the locals' rage was both a source of controversy and, paradoxically, a validation of their cause. Prominent environmental activists and scientists sympathetic to the campaign noted local activists' rage when criticizing certain tactics—especially, the use of violence—or the failure of the campaign's claims to square with science. *Vikāram* was the term used to describe the shortcomings of the *Jananīti* study, which "felt for the people." Similarly, Asha was a prominent environmental scientist who had long supported the campaign. Nonetheless, she said that she could not lend her voice to the campaign's call to close down the factory completely, which she believed was not justified by the scientific studies. She did not fault campaign participants for making this demand, which she said was grounded in their rage (*vikāram*). Given the people's experience of the pollution in Manamur, she explained, such emotions had arisen inevitably and spontaneously. But as a scientist, she held that she was "a person that believes in facing all of the issues with facts and figures." Thus, she declined to join them in acting based on emotion.

As in Satish's comments, Asha suggests that rage springs directly from local residents' long experience of suffering and struggle, making them volatile and irrational. This can, to some extent, make them act in objectionable ways—whether by fighting factory workers, threatening researchers, or making unscientific claims. Yet it can also serve as justification for these actions: they are out of control and cannot be blamed. Their outbursts can be read as an outcome of the dire circumstances which they have so long endured. Here, as in the pollution tours, there is an assertion of immediacy: the injustice of the pollution stirs up the *vikāram* of the people, driving them to rise up in protest. Their actions cannot be evaluated in the usual ways because they are the result of injustice itself, rather than of human reflection and agency.

Unlike the threats of violence with which political parties empty the roads, the semiotic power of uncontrollable rage, for Action Council activists, did not lie in forcing their opponents to submit. Rather, their outbursts of violence operated with a similar logic to their hunger strikes. In flashes of rage, as in displays of suffering, the protestors presented their actions as a direct, unmediated effect of the harm done to them. While displays of suffering in *satyagraha* are afforded by the vulnerabilities of bodies to harm, performances of rage play off another limit: the possibility of losing control. Fistfights and thrown stones were organic outbreaks of popular sentiment, not the organized machinations of political factions.

Displays of uncontrollable rage arguably takes this logic of immediacy to its ultimate extreme. In the grip of *vikāram,* the Action Council members' actions are pure reaction—bodily effects of the encounter with pollution. In the stance triangle's terms, they are not subjects but objects. They do not take moral positions with which one might align; they are integral, material components of the reality that others must evaluate. But what is true of those overcome by *vikāram* is also true, to some degree, of displays of people's protest more generally. These displays are not moral arguments; they do not analyze or deliberate. Rather, they seek to persuade by reliably representing injustices to the broader public. And the most reliable representations are not representations at all, but effects.

Analyzing similar discourses of spontaneity in the American Civil Rights Movement, Polletta has argued that denials of agency by social movement actors can encourage participation and heighten morale.[43] The rhetoric of *vikāram* points to one possible reason for this seemingly counterintuitive finding. The less reflective or strategic the actions, the more they point back to the reality of the pollution that drives them. The actions of the people are justified by the reality from which they spontaneously arise, and the existence of that reality is certified by the rage of the people. Likewise, members of the Manamur Action Council described *vikāram* as integral to their experience of participation in the campaign. It was to their *vikāram* that they attributed their campaign's ability to endure where other campaigns had quit or been bought out.

Yet there was also ambivalence about the loss of control that such passion entailed. In the aftermath of the quarrel with the TV reporter, the leaders of the Action Council chastised many of those involved. They were concerned that the reporter and his cameraman might portray the conflict in an unflattering way—perhaps bolstering the gelatin company's claims that the activists were not really "the people" at all but simply a bunch of hired thugs. Being an effect required careful management, and the Action Council's relatively small influence over the mass mediation of their performances of rage meant that flipping "out of control" was always a risky prospect.

## MORAL OBJECTS AND MORAL SUBJECTS

Ahmed and I waited for a good hour while Vijayan looked over the video we had taken. We were not eager for him to reach the end. Our footage stopped abruptly where the large protestor had charged at Ahmed, shouting at him to turn the camera off. It had been a tense moment, and one we had not fully understood. Had the Action Council come to see our camera, too, as a threat akin to that of the journalists from Kerala TV?

But if he saw this part of the footage, Vijayan made no comment about it. When he emerged from the other room, he seemed quite pleased. The footage showed, he said, that the reporter had acted very badly. Whatever the Action Council members had done, it was only a reaction to his provocations. He asked if we could compile some of the footage onto a CD for him to take with him. With much relief, we did so.

For the Action Council, people's protest was a reactive politics. While displays of violence in the street might seem incompatible with displays of fasting in the protest tent, both were ultimately displays of the effects of injustice. Like the guiding fingers at the edge of the paddy field, they were meant to direct attention away from the protestors' own words and actions, back to the pollution that suffused their everyday lives. Being the people was about making this link automatic.

In one sense, then, the environmental activists and the Action Council were alike. Both sought to distinguish moral subjects from objects and give priority to the latter. However, while for the environmental activists the aim was to purify their ethical positions of influence by other subjects, the Action Council aimed to become an object that would be evaluated sympathetically. The environmental activists distinguished themselves from the community-minded "common people" by their breadth of moral concern (see chapter 3). But the Action Council, in seeking support from outside their village, did not attempt to become more broad-minded themselves, nor even seek to appear so to others. Rather, they sought to position themselves as ideal objects of sympathy by depicting their actions as organic, unmediated reactions to their circumstances. Even as they sought to erase their own ethical subjectivity, the Manamur protestors' appeals situated these audiences as ethical subjects capable of sympathizing with distant others and taking action on their behalf—that is, as those concerned with the welfare of people beyond their own communities.

From this angle, we get additional insight into what went wrong in the collaboration between the radical environmental activists and the Action Council members (see chapter 2). The radical environmental activists went to Manamur looking to shape ethical subjects, hoping to transform Action Council members' values to align with their own and to guide the campaign in living out those values. But the Action Council's efforts to be the people had nothing to do with taking on new, more broad-minded values. On the contrary, their effort to be suitable moral objects was more consistent with person-centric efforts to save their village.

Yet we can also see in this why people's protests often *are* collaborations between local Action Councils and environmental activists working for broader social change. For their work to be meaningful, the radical environmental activists needed an object around which to align. More specifically, they needed a cause beyond their own community circles. As we saw in chapter 1, solidarity with people's protests filled this role. The Manamur Action Council sought to portray itself as "the whole people" in order to win support from beyond their village, and the radical environmental activists sought to fight for the whole people in order to break with the politics of identity groups, castes, factions, and parties. The differences between the two groups could be complementary, but only so long as the solidarity activists treated the Action Council as an ideal moral object and did not try to make it an ideal subject as well.

Despite their similarities, the two approaches also raise very different dilemmas. For the environmental activists, putting values over community could be isolating (chapter 3). The Action Council members' approach may seem to avoid that problem. But they had a different problem. They had to appear to present an entirely open, unmediated display of injustice even as they worked very hard to curate this display. All of the Action Council's efforts to present themselves as the people entailed risk of misconstrual. In each case, immediacy was largely a matter of interpretation. Thus, their pollution tours hinged on their visitors reading them not as those with merely something to say (as for the government researcher), but as those with something to show (as for Ahmed and the District Collector). Likewise, in displays of sacrifice, the Manamur Action Council relied on its hunger strikes being interpreted not as a carefully staged performance of suffering, but as a willingness to make the ultimate sacrifice. And in *vikāram*—where they might seem to be most like hired thugs, who are imagined as all force and no persuasion—their street violence was only powerful to the extent that it was seen to spring not from themselves, but from the injustices done to them by others.

The Action Council members' paradox can, to some extent, be seen in the structure of the triangle. An object only has moral valences to the extent that subjects take positions on it. But the Action Council needs to position subjects (literally, by bringing people to the tent, the paddy field, the edge of the river) while also erasing as much as possible their role in bringing about this positioning. Being a sympathetic moral object thus required constant management—not only to conduct pollution tours, scientific studies, hunger strikes, and outbursts of rage, but to ensure that the tours did not seem overly preachy, the studies not overly emotional, the hunger strikes not overly staged, and the rage not overly tactical. This could be very tricky work.

Yet being a moral object also held a special kind of power—a power to which even the most eminent political subjects must sometimes, perforce, concede. Although V.S. Achuthanandan's party members successfully steered him toward their great red tent, he never actually made it there. As he approached it, he slowed, finally stopping and refusing to take the stage. Standing in the street, he gave a speech about the need for unity among the people. The next day, much to the delight of the Manamur Action Council, this was the story covered by the daily newspapers (all of them, that is, except the CPM paper): that the revered politician had chided his own party for not uniting with the existing people's protest in Manamur. And it is not difficult to surmise why he had done this. His own reputation as a great leader depended on being seen not only as a party leader, but as a leader of the people.

This was what the Manamur Action Council had been aiming for: not to transform their own subjectivity but to become an object that attracts powerful subjects and enrolls their support. This time, it had worked.

It is tempting to end the story of the Manamur Action Council there—on a high note. This moment of mass media recognition was as clear a victory as any I witnessed during my fieldwork in Manamur. But of course, the struggle did not end that day. There were more marches and fasts, more scientific studies, more visits from renowned persons, and more fights with factory workers. There were even a few months when the factory ceased production and, as described in chapter 2, it looked like it might be moved elsewhere. But the factory was not moved. When I last visited Manamur—a full decade after my first fieldwork there—trucks were going in and out of the factory gates. The protest tent, meanwhile, was sitting empty. Vijayan said the protest was still active, but I no longer sensed the same energy—in him or in the air—that had been so palpable in those initial months after the police baton-charged protestors at the factory gates. Continuously performing the people is powerful because of the sacrifice it requires, the bodily commitment. For the same reasons, it is not easy to sustain.

Is this low note, then, the real end of the story? That remains to be seen. So go many stories of people's protests: like most of the stories we live through, their endings are often uncertain and evolving. But unlike many other parts of life, their beginnings and middles are so luminously purposeful, so compellingly focused on a particular end, that they lead us to want clear, final, victorious endings. In Manamur, such an ending was invoked every time protestors marched through the streets, chanting that the pollution must stop and the factory must be shut down. The same ending was printed on their signs and recited in their TV interviews, not to mention their interviews with me. Yet these imagined endings were always postponed, always in need of reimagining. Imminent victory slipped away partly because of the double-talk of politicians and the stalling tactics of government agencies, partly because of the sheer difficulty of sustaining protest. But there would most likely have been no victorious ending even if the gelatin factory had shut down forever. At least, not anytime soon. There would have been a fight over who should clean up the pollution. There would have been compensation to demand and medical bills to pay. So run the never-ending stories of some other famous people's protests.[44]

Against this background, the current lull in protest in Manamur is no more an ending than was the day that the CPM leader stopped between the two tents. The story's ending still awaits its next reimagining. This can be heard in Vijayan's insistence that the protest is still active (there is an ongoing legal case, he told me recently, that may be on the verge of a breakthrough), but also in assurances from solidarity activists that, so long as a factory is polluting in Manamur, the people will inevitably arise to fight.[45] As I write this, no one is marching for Manamur, but this does not mean that no one will ever march again. And in the meantime, the struggle to tell the story of a people's protest, victorious ending and all—continues on.

# Conclusion

## *Life Beyond Activism*

On Ahmed's last night with us, we all had questions for him.

The official mood was festive. For almost a year, he and I had been living with Adarsh, Faiza, and their daughter Tara, sharing meals, household chores, and all the mundane particulars that come with family life. Now, he was heading off to pursue an MSW at the Tata Institute of Social Sciences, home of the nation's premier social work program. To mark the occasion, I had brought home a feast of takeout: chicken, fish, mussels, fried rice, and various sweet treats. But as the evening wore on, a second mood began to surface, more inquisitive than celebratory. Sitting on the tile floor, cross-legged or propped on our wrists among the greasy plates and half-empty cartons, Adarsh, Faiza, and I began gently, with a note of humor, to pepper Ahmed with questions that we had never asked so directly before.

This mode of conversation was not altogether unusual in our house, and not only because of perpetual interrogation from the resident anthropologist. Ours was a house where questions were bluntly put and boldly answered. We regularly received visitors with strong opinions, and a couple of our most frequent guests were with us that night, as were Faiza's father and sister, who were never shy about speaking their minds. Yet Ahmed had always been conspicuously quiet in our debates. Over time, we had come to recognize that he, a devout Muslim, had some strong opinions of his own, but he had always tended to keep them to himself. Perhaps for this reason, now that he was on the verge of saying goodbye, we were all determined to pry these opinions from his lips.

What had been his first impression of me? Of Adarsh and Faiza? In his opinion, what were our bad traits? The more he hesitated and hedged, the more probing and provocative our questions became.

"And what about that night during Ramadan when John ate the food we put out for you?" Adarsh quipped, "How did that feel?[1]"

"If that food helped to quell John's hunger, then I am content," said Ahmed with a mischievous smile, refusing to take the bait. Adarsh howled, and laughter erupted all around.

Then I asked a question that quieted the laughter, one that made Ahmed take a long pause: when, during our year together, had Ahmed felt most afraid?

As I asked this, Ahmed's anxieties about his uncle and the quarry protests (see the introduction) loomed large in my mind. By this time, the Dialogue Journey was several weeks behind us, but the conflicts it had raised for Ahmed had never really been resolved. I also thought about the quarrel between local activists and the reporter on the bridge in Manamur—one local activist had charged at Ahmed, and nearly struck him, for video-recording this interaction at my request. After that incident, Ahmed had refused to return to Manamur for several weeks. But the story he told was not one I had ever heard. Nor was it at all the sort of story I had expected.

Ahmed told us about a day when he had purchased some vegetables on his way home. Faiza did nearly all of the cooking for our house, with occasional help from Adarsh, so Ahmed and I considered it our part to contribute to the raw materials, so to speak. That day he carried three plastic bags of produce across the soccer field that lay between the vegetable shop and our house. But as he got closer, he began to get anxious about how Faiza might react if he returned with these three bags. Not just anxious, he said, but afraid. He put everything down and carefully packed all of the vegetables into just two bags. Then he threw the third bag into a small water reservoir next to the soccer field, a place used by many of our neighbors as a makeshift landfill.

With this story, laughter regained the floor. Faiza teased Ahmed for worrying so much about such a little thing. What, she asked, could he have possibly thought she would *do* to him? Her sister needled Ahmed for finding Faiza so terrifying. Ahmed grinned sheepishly and chuckled, and I laughed too, even though I also felt sure that Ahmed had not meant this story as a joke.

I had had similar moments during my time among Kerala's environmental activists—slipping into the bathroom to take my allergy medicine or looking over my shoulder as I ducked into an ice cream store. I knew the anxiety he was talking about. It made one do silly things, but it was not a joke.

Nor, despite their laughter, did Faiza and Adarsh take Ahmed's story as mere comedy. Part of this was because they did not really regard the use of plastic bags, like such other "ordinary" things as tea or VapoRub, as a trivial matter. Like so many other targets of environmentalist concern, the mundane ubiquity of plastic bags—that they were part of the background for ordinary Malayali lives—only made them more problematic. As part of her work with a small environmentalist NGO, Faiza showed schoolchildren films about cities smothered in plastic bags. Every morning, Adarsh was careful to pack an empty fabric bag in his knapsack, just in case he needed to buy anything on the way home. But more than their concern about plastic bags as such, Faiza and Adarsh also heard Ahmed's story as a tragedy in another way: as a story about his failure to become an activist.

This was the same night, described in the book's introduction, when Adarsh and Faiza admonished Ahmed for remaining "normal" and "neutral" despite all he had experienced over the previous year. As Faiza told me later, it was Ahmed's story about the bags that had provoked this intervention. She saw his trouble with plastic bags as consistent with an unsettling pattern in his life: he was open to learning about environmental and social issues, but he was too concerned about what others thought of him, too hesitant to take a stand. Faiza believed that this was why, despite learning so much from his time with environmental activists—especially, despite his intimate acquaintance with Faiza and Adarsh's way of life—Ahmed had not become an activist himself. Anything he had learned from them had been outdone, she said, by his inclination to blend into any social context in which he found himself.

"I think maybe he has changed in his thinking. Maybe from now on when he goes to take a plastic bag, he'll think 'Oh, if Faiza was here I would not take this bag.' And then, chances are, he'll take the bag. Understand? His style is like that—all dependent on his context."

In her concern about Ahmed, Faiza articulated an idea that has returned in many forms throughout this book: being an activist is about putting values over relationships, commitment to a cause over the desire to belong. For Faiza and other radical environmentalists, this idea was fundamental to the struggle for environmental justice. Not all radical environmentalists lived this out in the same ways or to the same degree. Nonetheless, working for change was premised on a basic willingness not to blend in, a willingness to take a stand and face the social fallout. Some degree of conflict, some risk of social isolation, was essential to activism. More basically, it was essential to a good life.

In Faiza's speech to the art students at the nature camp, she had laid out a choice between two life paths—that of the common people, who take their values from their families, castes, parties, or religions, and that of the activists, who leave these communities in search of a better world. She seemed to sketch a boundary not only to environmentalist forms of life, but also to possibilities for the form of a fully human life. Likewise, in expressing her concern about Ahmed, she insisted that it was not simply a matter of whether he shared the same principles and opinions as she and Adarsh. Rather, her concern was that his life was not organized by any principles, that he was simply ruled by his relationships, his actions wholly "dependent on his context." Putting cause over community was the starting point for answering the question of how one—how *anyone*—should live.

What should we make of this idea? Even as it has recurred in the preceding chapters, I have also shown that there are other notions about how to live. Radical environmental activists in Kerala acknowledge this when they label their own way of life "alternative" (*badal*); their approach to values is defined by a contrast with the more prevalent morality of the "common people" (*sādhāraṇakkār*). Moreover, even among activists, theirs is not the only way. Local activists in Manamur have also organized their lives around the struggle for environmental justice, yet they ground their activism in their commitment to their communities, not in any quest for moral liberation nor any aspiration to attain justice for all. Not only is this a viable approach to activism, it is one on which the radical environmentalists' notion of "solidarity" depends. Moreover, even those most determined to live by principles alone—who, like Hari, are willing to stand, and stand, and stand, until they are standing all alone—may make us wonder whether such a life is the best way to advance the causes it proclaims.

Nonetheless, the choice that Faiza laid before the art students at the beginning of this book reasserts itself in all of the stories told here, even in the efforts of Manamur activists to fight for their own people. It lurks around Ahmed's story especially: not only because of Faiza and Adarsh's disappointment with him, but because he himself seemed to struggle with such a choice. As he marched with the quarry protest toward his home, as he returned from his first visit to Manamur full of new conviction, or as he stopped in the soccer field to reconsider the plastic bags in his hands, he seemed to teeter uncertainly between two possibilities for what his time among environmental justice activists could mean. As Faiza suggested, there was no question that Ahmed's time among activists had changed him. But how had it changed him? Had he gained new consciousness? Or had he only learned to fit in with a new group?

In asking about Ahmed's path, I have taken up Faiza's questions about him—her framing of what activist life, and human life, are all about. But my aim is not to make her life the standard against which Ahmed's is measured. Rather, I want to re-examine the assumptions that underpin Faiza's doubts, to ask whether these assumptions are appropriate to understanding Ahmed, and, from there, to probe the limits of Faiza's framing for understanding any human life, including my own. The story of Ahmed's time among activists offers a different angle on how people contend with tensions between *living for* and *living from*—both in struggles for environmental justice and beyond.

## ASYMMETRICAL INTERDEPENDENCE

In the preceding chapters, I used Du Bois's diagram of the stance triangle to describe how activism produces tensions between community belonging and fighting for a cause. While the triangle does not reflect the full import of these tensions, it aptly captures something basic to ethics: every act of evaluation is always also a way of relating to others. This speaks powerfully to the experience of activists like Hari, who persistently attempt to free their lives from concerns about group belonging yet find this difficult to accomplish. To some extent, the stance triangle illustrates why such difficulties arise: the evaluative (subject-object) and relational (subject-subject) legs of the stance triangle are inherently joined. But the triangle also provides a helpful framework for describing the aspirations that motivate these efforts. In the two diagrams below, I have adapted the stance triangle to describe Faiza's choice between the mainstream ethics of the "common people" and the alternative path of activists.

Both diagrams maintain the same basic structure of Du Bois's stance triangle, representing the intrinsic connection between the evaluative and relational aspects of ethics, between taking a stand on something and finding one's footing in relation to others. But in each of these triangles the interdependency of the original stance triangle is reimagined as

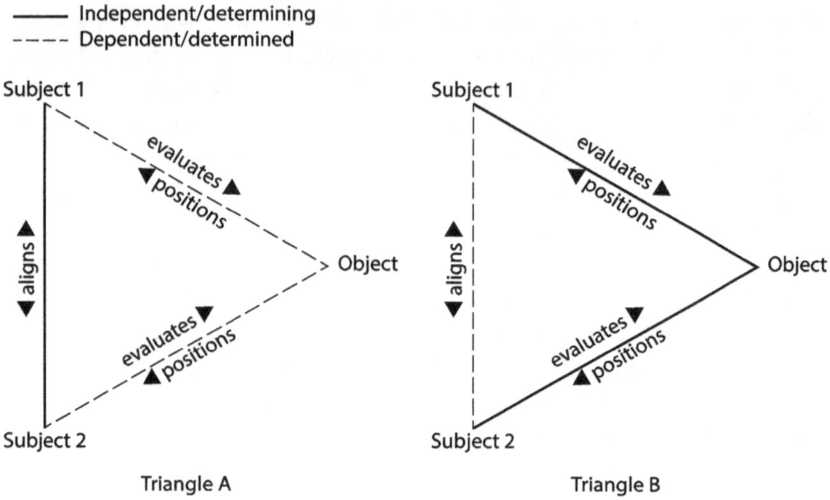


FIGURE 21. Faiza's choice in two triangles.

dependency—one aspect of ethics dominates, determining the other. Or, less absolutely, we might see these two triangles as introducing an asymmetry into the interdependency of values and relationships. Each needs the other, but one takes the lead role. In Triangle A, representing Faiza's account of the usual approach to values in Kerala (the approach of "ordinary people," *sādhāraṇakkār*), relationships take precedence; on the basis of the communities to which one belongs, one adopts various values and opinions. Living for begins by living from. In Triangle B, by contrast, values are independent of social influence and, moreover, common cause becomes the basis for building new communities; living from begins by living for. This is ethics as the radical environmentalists believe it should be. This is an activist ethics.

But is Triangle B *only* a description of activist ethics? Some might argue that this is the basic structure of any coherent ethics. For example, some of the most influential traditions of moral philosophy have sought to distinguish ethics from appeals to community norms, looking instead to universal moral principles, impartial spectators, and calculations of the greatest good for all.[2] Anthropologists, who seek to understand cultural variability in ethics, have often criticized this emphasis on impartiality, aligning instead with philosophical traditions—such as virtue ethics—that stress the importance of social roles and relationships to ethical life.[3] Yet a similar preference for Triangle B can be seen in anthropologists' efforts to draw clear boundaries around ethics as a domain of freedom and reflexivity, separate from the socially determined norms of morality.[4] In each case, the notion is that the very possibility of ethics presumes one's path will not mainly be a function of the communities to which one belongs but, rather, will be directed by one's own inquiry into good and bad, right and wrong. By such an account, the asymmetry between values and relationships in Triangle B is not simply one idea about how ethics can be done. It defines the limits of what can count as ethics.

This hierarchy between the two triangles may, in practice, be bolstered by its perceived implications for the scope of moral concern. For radical environmental activists in Kerala, preferring Triangle B over Triangle A was both a practice of freedom and a project of moral extension. With rain camps, a magazine, and other tools of moral inquiry, they cultivated a life "for the people" by subordinating community ties to principles, relationships to values. In this, they were well in tune with narratives of moral progress associated with modernity, especially the notion of environmentalism as the endpoint of a historical broadening of moral concern. A parallel asymmetry is evident in criticism of an alleged overemphasis on group identities in India, especially on caste and religion, which have been claimed to hinder Indian politics from measuring up to Western political ideals of nation, public sphere, civil society, or democracy—ideals grounded in the same tendency to see Triangle B as the foundation of ethics. This is not to say that Faiza's activism was somehow pro-Western or anti-Indian. But when she and other radical environmental activists undertake to live out this moral project, their lives speak, from within India, to the strengths and limitations of widely circulating notions about what makes a good person or a modern democracy.

This book unsettles this hierarchy, but it also tracks its persistence in the lives of Kerala activists. Environmental justice protests, insofar as they are understood to begin from concern for one's own people, might be seen as reasserting the place of community belonging in ethics and politics. Local activists in Manamur, in particular, seem to exemplify this, offering a counterpoint to the radicals' extensionist program for an environmentalist people's politics. And yet, the radicals folded local activism in Manamur back into their story about themselves—a story of solidarity that made protest of the gelatin factory an opportunity for transcendent impartiality and, in the same move, rendered the Action Council's person-centric aims invalid as a moral purpose. More to the point, even though the locals rejected the "broader" causes of the radicals, they ultimately took up their own version of object-oriented ethics in their efforts to persuade audiences beyond Manamur. The success of their campaign, they believed, depended on the ability of these audiences to take an expansive moral view, to care not only about their own people but about *the* people. Thus, even though the Manamur protestors seemed, at times, firmly planted in the world of Triangle A, their stories also seem to affirm the persuasive power of Triangle B.

Thus, the question remains: can one live a good life without being an activist? Listening to Faiza speculate about how Ahmed had or had not changed, I was acutely aware of this question—and of all the questions it seemed to thrust upon me. If our purposes are "dependent on context," do they really count as purposes at all? Or is it only when we pursue our purposes despite social pressures, or to the extent that we do so consistently, that our lives offer meaningful answers to the question of how one should live? What if Triangle A and Triangle B are not really two different approaches to ethics, two alternative visions for a good life? What if the difference between them is really the difference between a life with moral purposes, values, and principles, and a life that is missing something essential to ethics—courage, perhaps, or conviction—such that Triangle-A lives lack an ethical dimension altogether? What if an activist life was not only more "right" or "good" by some ethical standard, but simply a more fully human life?

And what, then, of an anthropologist's life? When I began fieldwork in Kerala, I had planned to be an activist anthropologist. My prior activism in support of environmental justice movements had been a source of contacts and rapport with activists in Kerala, and I had planned to study their activism, in part, in order to become a better activist myself. But the longer I spent in fieldwork, the more I had come to feel like a cultural chameleon, shifting my skin as I slipped from scene to scene, studying the radicals one day, the locals the next, then on to the factory workers, the tea-shop patrons, and our neighbors across the street. In each setting, I adjusted in order to participate and understand, and that too was a kind of purpose, but it also meant that I was a creature of context. Was this life inherently less complete than the activist life I had once hoped for?

Of course, no life matches up precisely to Triangle A or Triangle B. I have already shown how the lives of both radical and local activists could not be contained by such dichotomous categories, even when people appeared to be drawing the contrasts themselves. In this sense, Ahmed's life was the same: even though he described the incident with the plastic bags as a moment of fear, he was not a person driven purely by anxiety over social censure and exclusion. But one thing that separated both Ahmed and me from the activists we studied is that neither of us organized our lives so wholly around the daily pursuit of a cause. Did this mean that our lives were lacking by comparison? As I listened to Faiza, I wondered how the choice she had presented to the art students applied to his life or to mine. Inversely, I wondered what such less-activist lives might teach us about the interdependency of living *for* and living *from.*

## LIVING BETWEEN

Shortly before Ahmed and I joined the anti-quarry march that would pass near his uncle's home, we found ourselves stepping out of a small tour bus into one of the largest granite quarries in Kerala. We were taking part in a weekend visit to Faiza's ancestral home in the Western Ghats, a trip arranged by her father, which would culminate in a ceremony that he had organized in honor of her deceased mother. With us were Faiza, her sister, a few of their cousins, my wife, some neighbors from across the street, and a gaggle of kids of various lineages, including my own baby daughter. The first stop on this sight-seeing leg of our journey was the quarry.

As we walked up a steep hill, our sandals slipping in fine red sand, I wondered how our group had decided to stop here at all. The quarry had a brutal beauty; it sliced through the black rind of the rock face to reveal clean new granite, as white as bone breaking through the skin. But Faiza was leading our tour, and I felt sure that this could not be where she had wanted to bring us.

At the top of the first hill and the bottom of many more, we came to a flat place where dump trucks were parked and men reclined in the shade of a few wide-reaching trees. The trees ringed a metal shed, and inside the shed some machine made eating sounds—munch, munch, munch, munch—except it never stopped to swallow. On one side of the shed was a conveyor belt made of large scoop blades, like the buckets of an old-fashioned waterwheel, that lifted broken rock from an unseen cache below and dropped it into a metal box. A second conveyor belt extended from the front of the box like a long metal tongue, and from the tongue's tip an unbroken stream of gray dust poured down into a waiting truck, sending up a cloud of fine mist as it fell, like steam rising over pouring tea.

I felt sickened by this scene. But I wondered whether I would have felt the same a few months before. Over the course of my research in Kerala, I had begun to notice differences in myself that were both sensory and evaluative at the same time. Some were of the sort that one would expect in any long-term visitor: my food began to feel incomplete without the flavor of coconut, an ingredient in nearly every Kerala dish. But other changes were matters of taste in a different sense. Kerala's many rubber plantations had been cool, tranquil groves when I first encountered them years before. Now they felt desolate—the places where rainforests had been cleared. Elaborate mansions, with their high concrete walls, which I might have previously admired, were now only the ruins of mountains. The activists' influence on my own ethical views was no doubt at the root of these changes in taste. In this sense, I understood them. But I did not like them. For the most part, they made my world less beautiful. Now that quarries, rubber plantations, and mansions were distasteful, I could not experience them in any other way.[5]

For me, our whole quarry tour was like that. We climbed until we could take in a wide vista, with endless hills and valleys stretching away below. Here and there, naked black mounds erupted like stone bubbles from rolling waves of green. My eye immediately fell on the little white bite marks that had begun to gnaw the mounds away. Some had been chewed down to pencil-point spires, as spindly and white as church steeples. But I could not see them like steeples. I could only think that no one would ever know the size and shape of the mounds that had once been there.

The more I explored this new revulsion in myself, the more I wondered if Ahmed felt it too. As we came to one crest, there was a twisting cylinder of stone with another boulder, a huge block, balanced delicately on top, like the head of a hammer. Ahmed joked about how someone must have forgotten it there, a joke that had come into my own mind at almost the same moment. Such synchrony was unremarkable in those days, near the end of our work together, when we had so many months behind us. But on that day, as I felt out the changes within myself, this small moment of consonance made me wonder whether Ahmed might also have changed in similar ways.

If he had, how would I know? By this point in our time together, I knew that quarries were very much on his mind. Only a week before, he had learned that the route of the Dialogue Journey, the fifty-day march through the Western Ghats to raise support for conservation policy, might pass near his home. Already, he had confided in me that the anti-quarry focus of this event was a source of some anxiety for him.

FIGURE 22. Faiza guides Ahmed and I, with neighbors and friends, on a tour of a large granite quarry.

But now Rajendran, the lead organizer of the event, had suggested that Ahmed could organize a welcome party among the local youth in his village. In the days since, I had noticed him brooding about this. More than once, he had told me he was feeling afraid. "It all depends on which way they go," he said, tracing the Journey's possible routes in the air with his hands. "There is a split in the road there. One road goes off this way to [a large town], and the other one curves off this way and goes right past my house. As long as they do not go down the road to my house, it is okay."

I knew that quarries were on Ahmed's mind, but what did he think about them? Where did he stand? From time to time, he had said things that suggested he was concerned about environmental degradation in Kerala's mountain rainforests. He had even suggested to me, once, that the Quran offered a foundation for such concern. But how could I be sure that he was not, as Faiza alleged he was prone to do, simply echoing the views of those around him? He was clearly not willing to let his uncle find out about his involvement in quarry protests, let alone to openly protest his uncle's business. Or was it, rather, that he was not willing to let the activists find out that he was not so bothered by quarries after all?

Though I watched closely, I could find little indication whether our sightseeing tour of the quarry disgusted Ahmed as it did me. He and Faiza's cousins were mostly caught up in posing and snapping photos of one another in front of the boulders. Faiza made her own position clear with a short social history of the place, describing how it had once been a tourist destination for very different reasons: the boulders had been dotted with small ponds, worn into the rock by millennia of rain, and the water had been thought to have healing powers. Now, she said, all of that had been destroyed. Ahmed had missed this narrative the first time, and I had pulled him aside so she could repeat it. But it was hard to tell what it meant to him. As we walked back, past the dust-making machine, he had rejoined Faiza's cousins, who were snapping photos all the way down the hill.

In the days following our quarry tour, I continued to wonder about how Ahmed might, or might not, have been changed by his time among activists. But I was reluctant to ask him about it directly. As his employer and mentor in ethnographic fieldwork, I had learned to be careful about too quickly expressing my evaluations of what we encountered together. As noted earlier, Ahmed was often quiet about his views. In our first months together, I had felt that he was particularly reluctant to express views that might differ from my own. Over time, he had begun to more frequently and directly disagree with me about, for example, the role of caste in Kerala's party politics or the gender dynamics at a meeting we had both attended. But with regard to quarries and other environmental issues, my sense was that he was not entirely sure what he thought. I sensed that any direct question might push him to answer in a particular way. So rather than prying for answers, I waited and watched.

As the scheduled date for the Dialogue Journey neared, I thought I saw some signs that Ahmed really was more actively and explicitly adopting environmentalist views, including opposition to quarrying. Despite his anxiety about the route, he seemed eager to participate—and not only in the guise of the anthropological "participant observer." On the first day of the fifty-day trek, when some locals at a snack shop asked Ahmed and me what we were doing, he explained the mission and activities of the Journey in the first-person plural ("*we* are marching to save the forests"), as if he also was a part of the activist group. When he was done, I felt the need to clarify—given my status as a White foreigner—that I was there as a researcher, not a participant. In the days that followed, Ahmed joined other participants in visiting the homes of

FIGURE 23. Loading broken granite into a truck at a quarry site worked by Ahmed's uncle's company.

local residents and, though he took some notes and audio recordings, he was also active in discussion, expressing many of the same views held by the activists. He talked to me about this process, explaining that it had made him feel more a part of things. As we journeyed on, people treated him less like my research assistant and more like one among the younger generation of activists. And he seemed to welcome this role.

Yet I had also known him in other roles. A few months before, I had stood beside Ahmed in the bottom of one of his uncle's quarries, the high whine of jackhammers ringing in my bones, the deeper pulse of the hydraulic breaker rattling my teeth. It had been my idea to come here. Ahmed had been eager to bring me to stay a week with his family—to introduce me to his mother, his brother, his friends—but this had not been part of his plan. As I flipped my *mundu* up above my knees and went to help load chunks of granite into a truck, Ahmed hung back.

Less than ten minutes later, a piece of rock sliced my palm as it left my hand. While Ahmed went for Band-Aids, I joined his uncle in the small strip of shade near the base of the granite wall. Between long drags on a cigarette, he told me again what he had told me before: if you are unfamiliar with this work, you will get cut.

"Now that I've been cut," I said, "I guess I'm a bit more familiar with it."

"Not familiar enough."

He showed me his hands. Each was covered with a pale, smooth callous. Not just the pads of his fingers or his palms, but the whole face of each hand. He told me that when my hands looked like that, I would not have to worry about cutting them on the stone. Then he showed me a scar on his forearm as well, a raised white line several inches long.

Throughout the day, other men at the quarry showed me their scars. They all had them: seams and ropes of pale flesh spiraling around their forearms, calves, and thighs. One man ran his finger along the metal rod that held together the bones of a crushed foot. He flicked the top of the foot.

"I can't feel anything in this one now," he said. Flick, flick. "Nothing."

Neither Ahmed nor I mentioned our concerns about the environmental impacts of quarrying that day—or on any of our days at his uncle's house. It was not simply a matter of not wanting to disagree. Ahmed's uncle knew that we were doing research on environmental movements, and he knew that quarrying was at the center of the biggest environmental controversies in Kerala at the time. When I interviewed him, he volunteered his own criticism of unregulated quarrying and said he supported some of the policies that the Dialogue Journey was meant to promote. If he was an agent of the "quarry mafias," the term activists commonly used when denouncing quarrying, he was not an unreflecting agent.[6] Nonetheless, for my part, I felt like offering my own views would be talking out of turn somehow—both inappropriate and unnecessary. What good could it have done?

In our time with Ahmed's uncle, I began to empathize with Ahmed's position—and especially, with his tendency toward silence both among activists and at home. Ahmed is more familiar with quarry work than I am—familiar enough, at least, to know his limits. But like mine, his hands are uncalloused. His skin is unscarred and his feet whole because, even though he knows quarry work, it was never his work. His uncle had always had other plans for him. And it is because of his uncle's callouses and scars, and those of the men who labor for him, that Ahmed could leave the quarries, go to college, get a job as a research assistant with an American anthropologist, and eventually gain admission to the top MSW program in the country.

If Ahmed belongs more among the quarry marchers than among the quarry workers, that is only because he is so indebted to the latter. And so, paradoxically, even though he could not be a quarry worker, he could not really join the marchers either. He seemed to be caught between these two worlds, trying to balance his roles in each and reconcile their different values, belonging to both but not inhabiting either with ease.

I empathized with this position because, in a different way, my fieldwork had left me caught in between as well. Unlike Ahmed, who struggled with belonging to both the radicals and the quarry workers at once, I did not really belong in either world. In hindsight, it seemed to me now that I had once found this social distance empowering. Before my fieldwork, when I had contributed to solidarity activism with Indian environmental justice movements, my position had afforded a detached, global perspective from which the cause of "the people" and "the environment" had seemed clear. But now this detachment felt like a supreme disadvantage, an embarrassment, a lack of ground to say anything at all. Now that I too was enmeshed in relationships—though never nearly so deeply enmeshed as Ahmed—it was harder to know where to stand or what to say. And so, at least while we were at Ahmed's uncle's house, I too was silent about quarries.

## IMAGINED DIALOGUES

Such silences are one way of keeping conflicts between living for and living from at bay. Yet they also seem to affirm Faiza's criticism of those who are too "dependent on context." Like the incident with the plastic bags, Ahmed's silence about quarries raised the question of whether and how ethical values—matters of good or bad, right or wrong—were relevant to his life. His silence around his uncle might be seen as enacting values of respect or loyalty. But his silence around Faiza, Adarsh, and other activists was more difficult to read in this way. His response to the dilemma over the route of the quarry march—to keep quiet and hope they chose another way—seemed mainly to reflect a fear of being found out as kin of the "quarry mafia," not concern about the environmental impacts of quarries. It seemed to suggest that what had changed about Ahmed was not the ethical import of quarries or plastic bags, but only his recognition of the possibility of censure.[7]

Nonetheless, I sensed that our time among activists had changed Ahmed more deeply—and not only because I felt such change in myself. In one of my final conversations with Ahmed, when I had run out of time to wait and watch, I asked him frankly about how his opinion on quarries compared with those espoused by the Dialogue Journey participants. His response was uncharacteristically immediate and comprehensive; it was clear that he had already given the question careful thought. While he agreed that the really big quarries should be shut down, he felt strongly that activists were wrong to oppose all quarries. He argued that smaller quarries, of the sort his uncle digs, were needed if people were going to build houses for themselves. He said that the radical environmentalists' proposal that people should build "alternative" (*badal*) houses out of earth was simply impractical.

This explicit description of Ahmed's views, in and of itself, could tell me little about how Ahmed may or may not have been affected by our time among activists. To be sure, it clarified where he stood, positioning him somewhere between the activists and the "quarry mafias" they opposed. But it did not tell me whether he was changing. Indeed, Ahmed's distinction between smaller and larger quarries was similar to an argument I had heard from his uncle. Perhaps it was his uncle's views, not those of the activists, that he was echoing. It was only a moment later, when we began to talk about his silence with regard to these views, that I began to see how the activists were affecting him.

"Have you ever shared these opinions with anyone?" I asked.

"Oh, no, if I said that, they would all turn against me!" he replied. "Who knows what would happen?"

But although Ahmed had not told any of the Dialogue Journey participants his opinions, he explained that he had imagined doing so.

"At the time of those discussions, I stand up like that and say, 'I am really a quarry person!'"

"You said that?!"

"No, no! I thought that. I imagined. If I said that sometime, what would they do?"

Ahmed imagined this scenario in two ways. Laughing, he speculated that they would kill him. But he also speculated about the reasons that they would give for why he was wrong. He said that they would propose "alternative things" (*badalāyiṭṭuḷḷa kāryaṅṅaḷ*). For example, he surmised that they would point out that many wealthy Malayalis have built houses that are just sitting empty.[8] They would suggest that people should just share those.

"But not everyone will be able to accept that," he said. "But then, they'll probably say, 'Selfishness. You're not looking after the future, you're just looking after your own interests.' That's what they'll say. For each matter, they'll have some argument."

In this way, Ahmed voiced the activists' objections to his views, responded to them, and allowed them the possibility of response in turn. He acknowledged that some of their points were difficult to refute. For example, he agreed that current quarrying could be a problem for the next generation. He raised this point in the imagined voice of the activists, and he let it stand. But in response to other points, he gave succinct counters. Over months of imagined dialogue, Ahmed had worked out nuanced arguments on both sides. Although he had carefully kept silent about his opinions, he had also been talking with the activists all along.

Ahmed's imagined dialogues confound dichotomies between internal or external, self and other. Though the notion of "internal dialogue" is useful here, this process is informed by Ahmed's participation in the activities and discussions he had observed. In voicing the opinions of activists, Ahmed mixed past-tense reports of what they had said in similar situations with future-looking hypotheticals about how they might reply if he said such and such. Thus, the dialogue he presented shifted ambiguously between experienced and imagined, between overheard speech and inner speech. Ahmed had internalized the activists not as models for his own behavior, but as interlocutors.[9]

Yet this imagined conversation is also radically disconnected from these interlocutors. It allows their voices in, but it forecloses any possibility of censure, any risk of becoming out of place. In some sense, then, it makes possible the very sort of freedom prized by the radical environmentalists: here, consideration of point and counterpoint can proceed unfettered by concerns about community belonging. But the condition of this freedom is silence. Thus, there is also no possibility that Ahmed's arguments, no matter how well crafted, will persuade others. His imagined debates show that he, too, grapples with tensions between values and relationships—as much as he might seem to prioritize the latter. But they might also be seen as affirming Faiza's view that, despite his yearlong immersion in the struggle for environmental justice, Ahmed would not become an activist after all.

## ACTIVISM AS ONE POSSIBILITY

In the end, Ahmed was lucky: the route of the Dialogue Journey skirted his uncle's house. It did pass close enough that Ahmed was recognized by some of his local friends—to whom he carefully explained that he was participating as a researcher, not a protestor. But by then all talk had ceased of Ahmed organizing a welcome party among these friends.

Ahmed told me he was not sure why this proposal had been dropped. Perhaps Rajendran, the organizer who had originally made this suggestion, had gotten the sense that Ahmed was not interested. Perhaps the party idea had simply been forgotten. Ahmed did not try to find out.

A few weeks later, the morning after our farewell feast, Ahmed took a train to Mumbai to join his new social work program. Inspired by his experiences with environmental justice movements, he specialized in Dalit and Tribal Studies and did an internship with an NGO focused on environmental governance. When he graduated, he took a job with this same organization, where he conducted community-engaged research projects on environmental interventions that impact indigenous populations. In this career, he found his own way of living for environmental justice. Most of us would probably call this the life of an activist.

Nonetheless, Ahmed's path continued to diverge from what Faiza would have considered an activist ethics. By making a career of environmental justice work, he largely circumvented conflicts with his family, his neighbors, or his faith. He was not eating only fruit or spending all his time at camps and protests, let alone marching through the streets of his home village. He was off in another state, making a respectable salary. More generally, his life was not defined, as Faiza's was, by choices between community and cause, living from and living for. When he reached the age to marry, his uncle found a suitable spouse from the same Muslim sect and caste, the daughter of a friend and business partner in his home village. The engagement ceremony and wedding took place in the traditional way and with the traditional guests from among family and close friends, though attendance was lighter due to the COVID-19 pandemic. Ahmed objected to any dowry, but his wife's family nonetheless adorned her with the traditional ten and a half *pavans*[10] of gold jewelry, and she brought this wealth with her into the marriage.

This book has, in the main, examined the change narratives of those who are already activists—whether retrospective stories of one's own transformation or prospective efforts to bring others to the cause. But in such narratives, it is hard to fully appreciate the possibility that one could take an entirely different path, that one could also *not* become an activist. Whether one wishes to call Ahmed an activist or not, his story points to this possibility. Or at least, it points to the fuzzy margins of activist ethics—to viable modes of ethics beyond activism, but also to how tensions between community and cause can bleed into modes of life far removed from activism proper.

Such fuzzy margins can also be found in every life described in this book. Faiza married someone who shared her cause, and she was divided from her mother because of it. But Faiza valued family. On the first anniversary of her mother's death, she wept as Adarsh recounted how they had taken newborn Tara to Faiza's mother on her deathbed, so she could hold her once before she died. That was also the only time I had known Faiza to pray; she had looked up a traditional Muslim prayer for the deceased on Google ("That is the first time I ever prayed in front of a computer," she joked), and she explained that this was the prayer her kin would be praying in her home village that day.

Other fuzzy margins can be found in the lives of activists like Sunitha, who was only trying to be a good mother; or Dhanya, who wanted to put the people first, or Hari, who stood fast until he was all alone. Even in cases where people commit themselves to living solely for the cause, there are limits. And in cases where people like Ahmed choose to be silent about their views, the debate may still go on within.

As a vision for change, environmental justice helps us to see conflicts between living for and living from in bright colors and solid lines: the mainstream versus the alternative; solidarity activists versus locals; consistency versus compromise; universal versus particular. But as a form of life, environmental justice also continually confounds these dichotomies. Even when living *for* and living *from* come into conflict, they cannot be purified of one another. Even when they are aligned, tensions arise. For some, living as an activist may mean emphasizing one of these aspects of ethics over another. But the challenge at the heart of ethics—for Faiza and Adarsh, as for Ahmed—is to contend with both at once.

# Appendix: Note on Methods

This book is based on more than three years of ethnographic fieldwork in Kerala, spread across seven visits between 2005 and 2018. During this time, I employed a wide range of methods, including participant observation, interviews, audio and video recordings of social interaction, archival research, and media analysis. I compiled a large amount of data, including some 1500 pages of fieldnotes, 220 interviews, and more recordings than I could possibly revisit when I returned home. The book draws selectively on this material, focusing on what I came to see as the central stakes and dilemmas faced by environmental justice activists. Here, I offer a summary account of the key relationships, methods, and materials that led me to this analysis.

My fieldwork can be divided into three phases: preliminary, primary, and follow-up. Preliminary fieldwork consisted of three short-term visits (between one and four months) from 2006 to 2009 and a year-long advanced Malayalam course in Thiruvananthapuram in 2010–11.[1] During the first, shorter visits, I studied Malayalam and broadly explored Kerala culture and politics. In 2008, I completed a four-month internship with an NGO closely tied to the fishworkers' movement, which was my first introduction to "people's protests" as a local genre of politics. During the 2010–11 Malayalam course, I began to focus on the field sites and people described in this book. It was during this visit that I first encountered *Kēraḷīyam* environmental magazine with its vibrant activist community, including the couple met throughout this book, Adarsh and Faiza. These activists introduced me to the local leaders of the campaign against the gelatin factory in Manamur. Thereafter, I returned twice to conduct full-time, immersive participant observation in these communities, first for two months in late 2012 and then for fifteen months from 2013–14. The bulk of the empirical material for the book comes from this fifteen-month visit, during which my assistant, Ahmed, and I lived with Faiza and Adarsh. I also made follow-up visits in 2017 and

2018, during which I presented my initial findings to the activists I had studied, discussed my analyses with them, conducted additional interviews, and collected supplementary materials such as documents, films, and photos.

Two unexpected life events shaped my primary fieldwork in ways that, though challenging, ultimately proved serendipitous. The first was the recurrence of a chronic pain condition affecting my hands, making it extremely difficult to write notes ("jottings")[2] during participant observation. This happened in 2012, just prior to primary fieldwork. When I began seeking a research assistant whom I could train to write field notes, a faculty member in a BSW program in Kerala recommended one of her students, Ahmed.

By the time I arrived in Kerala, my condition had improved to the point that I was able to write most of my own notes. But Ahmed had already been hired. He and I lived and worked together daily for almost a year. His notes, questions, and opinions offered an invaluable counterpoint to my own observations. On many days, we would return home to record what we called "fieldnote conversations"—discussions, going back to our written notes, of all the day's events. These recordings served as a second set of fieldnotes. I have also gone back to him frequently to get help with transcription, to try and recall the details of some event, or to get his views on my analyses of activists' lives—including his own.

The second unexpected life event was the birth of my daughter in December 2012. Six months later, I took her with me to Kerala while my wife remained in the United States. Adarsh and Faiza, whose daughter had been born three months before my own, offered to have us move in with them. We rented a three-bedroom house together: one room for them, one for me and my daughter, and one for Ahmed. Even with the eventual assistance of an *āyah* (a person hired to help with childcare), caring for an infant made fieldwork very difficult; I have few notes from this time.

After two and a half months, my wife came to take my daughter back to the United States. In the meantime, however, my daughter had opened up doors for my research. My unusual position as a lone man caring for a baby made me a comical figure in Kerala, but it also seemed to make me more approachable and to arouse the sympathy of local mothers, leading to conversations and friendships that I would never have had otherwise. Previously, as a lone White man, I had occasionally been invited home by other men, to sit with them in the front hall or dining room while their mothers and wives served food from the kitchen. But now the neighboring women invited me onto the veranda to pass my daughter around; chat about her diet, sleep, and bowel movements; and eventually quiz me about my peculiar activist housemates. Even more important, however, were the relationships I formed with these housemates. I remained with Adarsh and Faiza for the duration of my research. As with Ahmed, my daily observations of their lives eventually became key material for this book.

Several criteria factored into field site selection. I realized early on that my field sites should reflect the two kinds of activists I encountered in environmental justice movements: those directly impacted by an issue and those who fought in solidarity. The Manamur gelatin factory campaign was one of the most prominent movements in the news at the time, while *Kēraḷīyam* magazine was a hub for organizing solidarity efforts. The *Kēraḷīyam* office was only about an hour's bus ride from Manamur, making it possible to study both simultaneously. The most crucial criterion by far, however, was trust. Beginning with my first visit in 2006, Sunny, the editor of *Kēraḷīyam,* had welcomed and encouraged me, saying that the magazine would benefit from a critical perspective on its work—a view that surprised me at the time, but which I later came to understand as typical of these activists' broader commitments. In 2010, when he joined as assistant editor, Adarsh took the same stance. This quickly led to invitations and new relationships, including an introduction to Vijayan, the main leader of the Manamur campaign. Vijayan also welcomed me warmly and encouraged me to study the campaign.

It is difficult to say where these activists' trust came from. There were also those who asked how they were to know that I was not from the CIA. Given the long history of American efforts to subvert Communist movements globally, including in Kerala,[3] this question was not unexpected; but I had no good answer. Others simply avoided me for reasons I never knew. Perhaps those who trusted me felt that I was too naïve, too interested in mundane things, or too encumbered with an infant to be a spy; people sometimes made jokes to that effect. But I tend to think that the reception I received from people like Sunny and Vijayan was more a function of their own personalities and commitments than anything about me. And because a few key leaders welcomed me early on, many others eventually extended trust to me as well.

In addition to my main field sites at *Kēraḷīyam* and in Manamur, fieldwork took me to several other sites that were helpful for comparison. I visited the offices of other small magazines and interviewed their editors about their work; I followed the work of solidarity activists who were not part of the *Kēraḷīyam* crowd, including a Muslim youth group with a mission of supporting people's protests; and I spent time with local activists in several other environmental justice campaigns, including campaigns against a Coca-Cola factory, a dam, an airfield, and a municipal garbage dump. I also cultivated relationships with opponents of the campaigns I studied and with people whose lives never, so far as I knew, touched the politics of environmental justice at all. These other scenes and lives are not the focus of this book, but I could not have written the book without them. They gave me a sense of the broader cultural stakes in environmental justice activism. They also gave me glimpses of the life possibilities that the activists described here, whether eagerly or reluctantly, had set aside.

On a daily basis, most primary fieldwork was conducted in tandem with Ahmed. Participant observation dominated. This was conducted in the traditional manner; we followed activists wherever they went and took notes, photos, and audio and video recordings, which I used to write up fieldnotes in the evenings. In choosing whom to follow, I generally sought to maximize the range of different perspectives. But I also interviewed some people simply because they were central to a particular activity (such as the organizer of a series of protests). My approach to interviews built off of what I had seen in participant observation; for the most part, I waited to do interviews until I had spent enough time with a person to feel I knew the right questions to ask. Some people, like Faiza and Adarsh, I interviewed multiple times, to document their views as circumstances changed.

Video recording also came to play an unusually large role in my fieldwork, serving as a complement to handwritten notes during participant observation. Initially, this was motivated by a desire to conduct fine-grained semiotic analysis of social interaction, but it ultimately became a way of questioning the assumptions or interpretive frames baked into my field notes.[4] Because many activist gatherings sought to attract news media, my own recordings of such events were not generally seen as unusual. Sometimes, with permission of those present, I would record less public interactions as well. Generally, I would set up a video camera on a tripod in one corner, hit record, and leave it while I focused on taking notes. When following more diffuse or mobile interaction, such as a rally or a march, Ahmed would record video while I took notes. In either case, the resulting footage was rarely aesthetically pleasing, but I never intended to present it publicly; it was simply an additional record of what had happened. Occasionally, activists would request copies for their own purposes—for a web page, or to show friends an interaction they found interesting. Otherwise, the resulting repository of video allowed me to revisit and analyze details of interaction that my notes could not possibly capture in real time. These details enriched my vignettes and, at times, afforded analysis of the nuances of language use, gesture, or bodily positioning.

Archival work was also important to fieldwork, particularly as I sought to understand the historical roots of "people's protests" in Kerala. This work began at the Appan Thampuran library in Thrissur, which holds a public archive of early print media including several small-circulation magazines. For many small magazines, however, I was forced to assemble my own archive, visiting older activists and digitizing issues from their private collections. For several months, I also collected daily newspapers and catalogued all appearances of coverage of the gelatin factory protest in Manamur. To a more limited degree, I tracked coverage of other environmental justice issues as well and recorded relevant TV news broadcasts. All of this helped me to understand the role that mass media played in the politics of environmental justice as well as how movements like the Manamur campaign were represented in public discourse.

Data analysis began while I was still in the field. This was an iterative process of identifying apparent patterns and rough hypotheses and then questioning them in further fieldwork, asking where I might find perspectives or empirical materials that would challenge my emerging account. Conversations with Ahmed were crucial; after the first couple of months, we were already actively seeking out new material relevant to this book's central themes of belonging, alternativeness, evaluation, and evidence. This process also benefited from conversations with academics in Kerala, especially J. Devika and Nizar Ahmed, both of whom knew many of the activists I was studying as well as the scholarly conversations in which I hoped to intervene. Through all of these conversations in the field, I gained a rough sense not only of what I wanted to write about but also, more viscerally, why I would write.[5]

When I returned to the United States, I was faced with the challenge of organizing the overwhelming amount of data I had accumulated. I chose to focus on my fieldnotes, which were arranged chronologically by day. I rearranged all of my other materials (formerly organized by sites, persons, or events) into folders for every day as well, so that I could easily find the files that supplemented the notes. In multiple passes, I layered codes into my notes, using searchable tags (#Family, #Alternatives) in MSWord to track themes or lines of inquiry. For interviews and other recordings, I used ELAN linguistic annotation software to add layers of transcription and coding in a similar way.

Analysis also benefited greatly from dialogue with the activists I was studying. In 2014, at the conclusion of my primary fieldwork, *Kēraḷīyam* hosted a two-day seminar on the relation between social science and social change, in which I gave an extensive (though very preliminary) presentation of my work and discussed it with activists, including several key activists from Manamur. I gave updated results at a second one-day seminar in 2016. Both events were held in Malayalam. Attendees at these events gave me excellent critical feedback that shaped further analysis. Later, I circulated drafts with Adarsh and others. Ahmed has also given frequent critical feedback, including on the final versions of the manuscript.

# Notes

INTRODUCTION

1. To protect confidentiality, I use pseudonyms for people and for some places.

2. Throughout the book, I refer to these activists primarily as "radical environmental activists," or simply as "radicals." These activists did not often refer to themselves as a group, a pattern consistent with their opposition to social conformity and bounded communities (see chapter 1), and there was no obvious local term to describe their type of activism. In this, they differed from those I call "local" activists in Manamur, who often referred to themselves as *nāṭṭukār,* a Malayalam term approximating "locals" (see note 13 below, and chapter 2). The locals in Manamur also had a word for the radicals— *pāristhitika pravarttakar*, "environmental activists"—but some of the radicals objected to this term on the grounds that not all of them shared a common environmentalist ideology or program. Key interlocutors suggested that "radical" was the best term, because it captured the commitment to broad change without specifying any particular ideology or program.

3. As the coming pages will show, the formulation of each of these categories of activism was part of the ongoing work of fighting for environmental justice. In this sense, they fit with Kim Fortun's notion of "enunciatory communities," which are produced by the "fields of force and contradiction" that generate environmental conflicts (2001, 11). The commonality within these types and the boundaries between them are not fixed; they respond to the strategic process of struggle. But this does not mean they are entirely fluid; patterns of protest (such as the circulation of a magazine or daily gatherings in a tent) contributed to their stabilization as types (see chapters 1 and 2).

4. Faiza had also led similar nature programs for mixed-gender groups. Generally, however, young men predominated at overnight programs, which par-

ents were less likely to let their daughters attend. Thus, even though the choice Faiza laid before the art students was her own, it was often less available to women than men (see chapter 3).

5. King 1997 [1963].

6. The term *sādhāraṇakkār* can also be glossed as "ordinary people," but it did not necessarily carry the negative, condescending overtones of either of these phrases in English, and people commonly used it to refer to themselves. Used in other contexts, to say something is *sādhāraṇa* could mean that it is lesser than something else; a *sādhāraṇa* jewel, for example, means a facsimile or manufactured stone—a jewel of lower value. And in most uses, *sādhāraṇakkār* held some of this meaning. It was commonly contrasted with elites—with politicians, intellectuals, or the wealthy, for example—each of whom had something (power, education, money) that the *sādhāraṇakkār* did not. Yet this did not necessarily mean these people were better than the *sādhāraṇakkār.* Rather, just as "the people" were valorized by way of a contrast with elites, so also for *sādhāraṇakkār.* The term could connote virtuous simplicity, humility, or a lack of corruption by worldliness. In this sense, the radicals' use of *sādhāraṇakkār* as a foil for their own ethical project was unusual. For them, to be common was to focus on status and wealth—not necessarily to be middle class, but to aspire to be middle class. The disjuncture between these notions of what it means to be "ordinary" may be rooted in deeper cultural tensions between socialist and consumerist ideals of progress in contemporary Kerala (Mathew 2022; Sunilraj 2023).

7. Questions about the scope of "moral standing" (Brennan 1984; Jaworska 2007), "moral considerability" (Goodpaster 1978), or "moral status" (Jaworska and Tannenbaum 2021; Warren 1997) have been seen by some as the core problematic of environmental ethics (Brennan and Lo 2002). Environmentalists often point to Aldo Leopold's (1949) "land ethic" as a key early conceptualization of the need to broaden moral status beyond humans. Later scholars have conceptualized environmentalism as the culmination of a progressive expansion of moral standing in Western thought (Nash 1989; Singer 1981). However, some have also seen Western thought as uniquely anthropocentric and looked elsewhere for inspiration (e.g., White 1967; Callicott 1987). For critical analysis of analogies with Indigenous and "Eastern" value systems, see Guha 1989; Whyte 2015.

Explicit historicizations of environmentalism can also be found in broader concerns with "speciesism" in environmental discourse. Kopnina (2014, 9) writes, "Just as one day slavery has become intolerable, perhaps one day the subordination of non-human species will become unacceptable." See also position pieces by animal rights activists (Ryder 2010; PETA n.d.) and an opinion piece challenging speciesism in social work (Wolf 2000).

8. Most proposals for moral extension, or the extension of moral status beyond humans (Engel 2008), can be categorized by the boundaries that they draw. Thus, extensionist environmental ethics may be zoocentric (extend status to all animal life), biocentric (all life), or ecocentric (all key ecosystem actors) (Batavia et al. 2020). Generally, these proposals call for gradations in moral consideration across this scope rather than for total equality. However, in such graduated schemes, moral preference may be calibrated not to species distinc-

tions but to such cross-species considerations as sentience or cognitive capacity, as in Singer's (2009) proposal.

9. Among numerous proposals in philosophy, statements from eco-feminism (for example, Plumwood 1993; but see Diehm 2010 on differences regarding the grounds of extension) and deep ecology (Næss 1973) have been especially influential. Such proposals are also being put into practice in professions like law, medicine, and social work. Lawyers and legal scholars are extending the legal category of "person" and the framework of human rights to nonhuman entities (Boyd 2017). Doctors and public health professionals are looking beyond human health to ecosystem health (Mackenzie and Jeggo 2019). Ecosocial workers are expanding social work's traditional mission of serving vulnerable human populations to serve nonhumans as well (Krings et al. 2018; Dominelli 2012).

10. Mason and Rigg 2019; Callicott 2013; Attfield 2013; Mathews 2013.

11. Some activists contrasted their "values" (*mūlyaṅṅaḷ*) with *dhārmmikata,* a term more often used in religious settings and associated with notions of purity, sin, and the conduct of one's proper role in the social order.

Dumont (1970, 251) defines *dharma,* the Sanskrit root of the Malayalam *dhārmmikata,* as "action conforming to universal order." It is, in part, the connection Dumont describes between this moral concept and social hierarchy—especially, though not exclusively, caste hierarchy—that makes this concept distasteful to the activists described here. This stands in contrast to some scholarship on Indian environmentalisms that has seen in *dharma* convergences between Hinduism and environmental ethics (Jain 2011). *Mūlyam,* the singular of *mūlyaṅṅaḷ,* can be used broadly to describe orders of value other than the ethical, such as economic value. It is this semantic breadth that, arguably, makes it attractive for appropriation and resignification within a variety of activist ethical projects—especially, as a way of disconnecting ethical evaluation from the moral orders of religion and caste.

In their study of ethics among queer activists in northern India, Dave (2012, 6) makes a similar distinction between "morality as norms of 'proper' gendered, sexual, and familial comportment" and "ethics as the *undoing* of social moralities." In an analysis that resonates with the self-understanding of radical environmental activists in Kerala, they argue that this makes activism "clearer as a kind of ethical practice, distinguished from moralities by its creatively oppositional relationship to the normalization of life and words."

12. I offer Malayalam glosses of these terms to indicate their common usage, and I look more closely at *paristhiti* and *janaṅṅaḷ* in chapters 1 and 2, respectively. However, English and Malayalam do not live in separate worlds. My inclusion of Malayalam terms should not be taken as signaling a "local" language of environmentalism, disconnected from "global" environmental discourse. Like many Malayalam speakers, radical activists mixed English terms such as "ecology" into their speech and used these to describe their cause. Many also read widely and were familiar with environmental thinkers and debates abroad; their own ideas were eclectic and wide ranging (see chapter 1). Thus, throughout the book, I am careful not to depict their activism as grounded in essentially Indian or "non-Western" concepts and values. Contrast this, for

example, with Vandana Shiva's (1988) argument that the environmental thinking of Indian peasants centers on a notion of *prakṛti,* a feminine principle of being that, she argues, differs fundamentally from Western conceptions of nature. Radical activists in Kerala discussed related conceptions of *prakṛti* and "nature," at times drawing similar contrasts between Indian and Western concepts (see chapter 3). Many admired Shiva's work; when she came to speak about her ideas, her talk was well attended. Yet these activists discussed Indian environmental ideas alongside those of foreign thinkers like Henry David Thoreau, Rachel Carson, or Masanobu Fukuoka; during my research a major event was the release of a Malayalam translation of *Ecology as Politics* by Austrian Marxist André Gorz. Thus, while their activism was clearly Indian, it is not particularly helpful to categorize it as either "Western" or "non-Western" (Venkatesan 2021; Nandy 1983).

13. This latter term is taken from the "solidarity committees" (*aikyadārḍhya samitikaḷ*) that radicals at times formed in support of people's protests, notably the committee supporting local protest of the gelatin factory in Manamur. The notions of "radical" and "solidarity," as I use them here, also share a common logic of detachment. As I describe in chapter 1, radical activists, by seeking to transcend community allegiances, set themselves apart from others. Likewise, as Fiona Wright (2016) argues, to work in solidarity is to ally one's efforts with a person or group while also taking them as other.

14. The Malayalam term I gloss here as "locals" (*nāṭṭukār*) is derived from the term I gloss as "our village" (*nāṭ*) in the preceding quote from Sunitha. *Nāṭ* could be more literally translated as "place where X belongs," with the meaning depending on who is X (cf. Daniel 1984). See chapter 2 for further discussion of this term and, more generally, the importance of local belonging to environmental justice movements.

15. Classic accounts of the emergence of environmental justice as a named, publicly recognized movement include Bullard 1990; Bryant 2003; Taylor 2014, 2000; Cole and Foster 2001; and McGurty 2009. As the movement's many historians note, however, communities of color in the United States had been fighting environmental racism for decades prior. Pulido (1996) describes such protest movements among Chicanos in the American Southwest (see also Perkins 2021), while Gilio-Whitaker (2019) traces the Indigenous struggle for environmental justice in the US to the very beginnings of settler colonialism and the oppression and displacement of Indigenous populations (see also La Duke 1999; Whyte 2018; Estes 2019). K. Smith (2021) shows how the environmental justice concept was influenced by deeper traditions of Black environmental thought.

16. Pezzullo and Sandler 2007; Checker 2011; McGurty 2009.

17. Pezzullo and Sandler 2007; Taylor 1993; Sze and London 2008; Pulido 1996; Gilio-Whitaker 2019. Some of the most powerful criticism of this tendency in predominantly white American environmentalisms has come from outside the United States. Guha (1989) offers an influential criticism from a "Third-World" perspective, drawing largely on his research with the Chipko movement against forest degradation in northern India (Guha 2000 [1989]). Martinez-Alier (2003) likewise writes of the "cult of wilderness" and the "gospel of eco-efficiency" that dominates American environmentalism, contrasting this with

environmental justice in the United States and the "environmentalism of the poor" in the "Third World," both of which integrate environmentalism with struggles for social justice (see also Guha and Martinez-Alier 2013).

However, Baviskar (1997, 2005, 2020b) argues that the category "environmentalism of the poor" obscures the dominant positions of urban, highly-educated, middle-class activists in northern Indian environmental movements. In studying collaborations between these activists and tribal communities that, in many ways, align with the collaborations between solidarity activists and locals described in this book, Baviskar shows that the middle-class activists "almost exclusively defined what sustainable development is and who is a fitting flame-carrier for the cause" (1997, 222). In research on urban environmentalism in Delhi, she describes the dominance of "bourgeois environmentalism, the (mainly) middle-class pursuit of order, hygiene and safety, and ecological conservation" (Baviskar 2020a, 329). These accounts point to the class diversity of Indian environmentalisms and show how the distinctions drawn by Guha and Martinez-Alier do not neatly align with national borders.

18. The most obvious example of how environmentalists' expansive ideals can turn against humans is the Voluntary Human Extinction Movement (VHEM), which argues that it is best for humanity, as the primary threat to nature, to come to an end (Ormrod 2011). Yet an anti-human potential is arguably present in all environmentalisms that see humans as a threat to nature, and this core antagonism can surface in various guises. A striking example can be found in the work of utilitarian philosopher Peter Singer, whose book *The Expanding Circle* (1981) crystallizes the case for moral expansion as a form of moral progress (see chapter 2). From utilitarian principles, Singer argues that moral standing should be calibrated to criteria like rationality and consciousness rather than species prejudice. On this basis, he has claimed that parents should have the right to kill newborn "defective infants," including those with such conditions as spina bifida and hemophilia (Singer 1979; see also discussion in Schaler 2009). Disability rights advocates have seen this argument as "bone-chilling" and an existential threat, and Singer was subsequently ostracized for his views (Johnson 2003). Yet Williams (2009), one of Singer's most cogent critics in philosophy, argues that his position is not inconsistent with broader criticism of anthropocentrism and speciesism, which Williams suggests begins from lofty ideals of impartiality but may ultimately tend toward "a hatred of humanity" (96).

19. I build on this global discourse when I refer to the activism described here as "environmental justice." While Kerala activists sometimes used this term to refer to their own work, the far more common Malayalam term was *janakīya samaraṅṅaḷ*, "people's protests." For discussion of the development and global spread of the environmental justice concept, or frame, see Taylor 2000; Schlosberg 2013; Martinez-Alier 2016; Martinez-Alier et al. 2016; Temper, Del Bene, and Martinez-Alier 2015. Related terms include "environmentalism of the poor" (Martinez-Alier 2003) and "subaltern environmental struggles" (Pulido 1996), but environmental justice has gradually emerged as an umbrella term. While initial definitions of environmental justice focused on inequity in the distribution of environmental harms and benefits, scholars and activists have since expanded definitions to include notions of procedural justice, recognition, and

capabilities—notions that reflect the concerns and strategies of environmental justice movements (Schlosberg 2007). Álvarez and Coolsaet (2020) argue that these definitions of environmental justice remain limited by their roots in Western movements and scholarship and fail to capture notions of justice in the global South. Gilio-Whitaker (2019) likewise points to the limits of the environmental justice concept for Indigenous activism. Pellow (2017) has called for critical environmental justice, which would extend the paradigm in more ecocentric directions. Elsewhere, I have argued that environmental justice is best understood as a living tradition (MacIntyre 1981), defined chiefly by a shared problem—the question of how to integrate social justice and environmentalism—rather than by a shared set of principles or practices (Mathias, Krings, and Teixeira 2023).

20. Nonetheless, some continue to question whether the localized, community-centric aims of activists like Sunitha are at odds with environmentalist values. The perceived conflicts are both conceptual and strategic. Proponents of ecojustice claim to improve on the environmental justice concept by advocating justice for nature as a whole, rather than only for humans (Washington et al. 2018; Kopnina 2016). In a more practical vein, Méndez (2020) describes how mainstream environmental organizations in California saw the environmental justice agenda of local Latinx communities as a threat to their climate policy agenda, which focused on reducing global carbon emissions.

21. This division of activist labor is not unique to Kerala, nor to environmental justice activism. Globally, environmental justice campaigns are commonly identified with particular places and proceed as collaborations between those who belong to those places and those from "outside" (Gardner 1995; Rootes 2007; Méndez 2020). Further afield, distinctions between "local leaders" and "organizers" in American community organizing traditions exhibit a similar structuring of roles (Alinsky 1971).

22. In *Economy and Society* (1968 [1922]), Weber argues that "the specific task of sociological analysis" is "the interpretation of action in terms of its subjective meaning" (8) and immediately goes on to claim that "processes and conditions . . . are devoid of meaning in so far as they cannot be related to an intended purpose" (9). Thus, while Weber recognizes the importance of other causal factors in human action, he places purposes at the center of the social analysis.

23. This notion is so foundational to anthropology, so baked in, that an "anthropology of purposes," as a specialized topic, would arguably be superfluous (but see Kavedžija 2016).

24. Some would argue, against this view, that people choose their own purposes; perhaps the choice Faiza presented to the art students could be understood in this way. Yet on what basis do people make such choices? A helpful discussion in philosophy centers on a moral dilemma that echoes Faiza's choice. Jean-Paul Sartre (2007 [1946]) recounts an experience from the life of one of his students: in the midst of World War II, a young man must choose between caring for his ailing mother and leaving home to join the French Resistance. Sartre argues that no moral system or set of values can inform such a choice, which is between "two kinds of morality: a morality motivated by sympathy and individual devotion, and another morality with a broader scope" (31). He argues

that one can only respond to such a dilemma with a radical choice, and that it is by such choices that we give value to our purposes. Responding to this, Charles Taylor (1976) asks whether such a choice, "made without regard to anything," can even be called a choice (293). He argues that such dilemmas may lead instead to deep personal reflection about "those inchoate evaluations which are sensed to be essential to our identity" (299; see also Descombes 2016; Schwenkler 2017). I thank John Schwenkler for pointing me toward this discussion.

25. Benedict (1934) gives a classic account of culture that fits this description. But criticism of the culture concept also reiterates this emphasis on locatedness, both of researchers and of those studied (Abu-Lughod 1991; Trouillot 2003).

26. The foil for this work is Émile Durkheim's moral theory, especially his cross-cultural study of religion, which stressed the centrality of community as both the source and the purpose of ethics. Durkheim (1915) argued both that our values come from the communities to which we belong and that the primary function of ethics is to bind these communities together. While this picture of morality is more fully developed in Durkheim's study of religion, it is already present in his doctoral dissertation, *The Division of Labour in Society* (1984 [1893], see, especially, 310–11). In the posthumously published *Moral Education* (1961 [1925]), Durkheim takes a programmatic approach to the same conception of morality, asking how social institutions can be employed to sustain morality (and, thus, society itself) in the absence of religious beliefs. Notably, in *Moral Education,* Durkheim also puts forth an account of moral freedom that, he argues, is only possible via social discipline and community belonging. This concern with freedom, if unsatisfactory to his critics among contemporary anthropologists, is arguably motivated by concerns similar to their own. For a discussion, see Zuckerman (2018).

27. Building on the neo-Aristotelian virtue ethics of Foucault's later work, Faubion (2011, 3–4) has described ethics as a process of reflexive self-fashioning, or "autopoesis"; actors may be incited to adhere to new norms or moral codes, but they are only ethical actors if they do so "freely and self-reflexively." Likewise, Laidlaw (2014) accuses an older, Durkheimian emphasis on the reproduction of social norms of missing the real action in ethical life by conflating morality with deterministic processes of social "unfreedom." As an alternative, he draws on Foucault to define ethics as "practices of freedom." Zigon (2008) makes a similar argument from a Heideggerian perspective, arguing that Durkheimians gave exclusive attention to unreflexive following of moral codes and missed the importance of self-reflexivity. Mattingly (2012) argues that Foucault's virtue ethics is of limited value for foregrounding first-person aspects of ethics such as "self, agency, experience, motive, self-interpretation" because Foucauldian analysis still defines these as "the *effects* of collective practices" (175, italics in original). She suggests that a humanist or "first-person" tradition of virtue ethics, building on Heidegger, Anscombe, MacIntyre, Arendt, Cavell, and others may be better suited to attend to "the problem of action itself, to the *doing* of ordinary life" (2014, 55 italics in original).

28. Laidlaw 1995.

29. Mahmood 2005.

30. Calls for attending to the social embeddedness of ethics span the philosophically diverse breadth of the anthropology of ethics. As Venkatesan (2023) notes, even the most ardent proposals for focusing on moral freedom recognize that freedom is always embedded in social relationships. For a Heideggerean argument in this vein, see Zigon (2021). For Foucauldian arguments, see Laidlaw (2002, 2014) and Faubion (2011). For discussions of the embeddedness of ethics in social interaction, see Lempert (2013); Keane (2016); and Sidnell, Meudec, and Lambek (2019).

My own approach cleaves closest to this last tradition in that it sees ethics as, fundamentally, a way of relating to others. In doing so, it builds on a lineage of moral thought that includes Adam Smith (2002 [1761]), who grounded ethical sensibility in "fellow-feeling" and the human propensity to see oneself through others' eyes. Other key influences from this lineage include George Herbert Mead's (1934) and Erving Goffman's (1959) respective theories of the self as a product of social interaction, Joan Tronto's (1993) ethics of care, and Judith Butler's (2005) writing on the social dynamics of accountability.

31. Faubion 2011, 85–86; Keane 2016, 17–20; for a review of arguments for and against distinguishing ethics from morality, see Mattingly and Throop 2018.

32. The contrast between *living for* and *living from*—between purposes and belonging, cause and community—shares some of the stakes found in distinctions between ethics and morality, but the prior contrast also differs in key ways. Distinctions between ethics and morality vary widely; for example, compare Williams (1985); Foucault (1990); and a discussion in Keane (2016). But they often depend on inherently opposed terms: freedom and unfreedom, critical reflection and unreflexive habitus, norm-breaking and norm following (Dave 2012; Zigon 2008; Laidlaw 2002, 2014). Moreover, these oppositions are often put forward as part of arguments for focusing anthropological study on one side or the other (but see Robbins 2007). Living *for* and living *from* are not opposed by definition, though they denote distinct aspects of social life that can often be in tension in practice. This opens the way to empirical study of many of the stakes one finds in distinctions between ethics and morality. What happens, for example, when people try to take some particular vision for moral freedom as far as they can? What happens when they make community belonging central to how they pursue their purposes? To keep things open-ended as I explore these questions, I do not make any analytic distinction between ethics and morality in this book.

33. Prasse-Freeman writes "Activists are hence guides to alternative futures, laying down different pathways that people could potentially take" (2023, 274). Anthropologists and other social scientists have often turned to activism as a domain for discovering new, more hopeful possibilities for social life (Graeber 2009; Juris 2008; Tsing 2005; Sitrin 2006). In such cases, scholars who study activists have often also become activists, joining those they study in fighting for a shared cause (Hale 2006; Scheper-Hughes 1995; Speed 2006). As a counterpoint, Shah (2010) warns against erasing the ways that activist projects may perpetuate injustice and argues that critical ethnographic analysis of the shortcomings of activist efforts can also lead to "ideas of how life could be lived, or imaginings of the world, which may show the potential for a radical politics that can better serve the poorest people" (190; see also Howe 2013,

169). In this book, I attempt to glean lessons both from the possibilities opened up by activists' efforts for change and by the limitations they confront in realizing those possibilities.

34. Harrison 2010 [1997], 2.

35. As activism has increasingly been recognized as complementary to inquiry in the social sciences, engaged methods have become the norm in studies of social movements. An enduring exception are studies of activists whose aims differ significantly from the researcher's own commitments (Powell 2020). Yet such studies have long been a minority in social movement studies (Edelman 2001).

36. Among many such calls, statements by philosopher Donna Haraway and sociologist Patricia Hill Collins have been especially influential. Haraway (1988, 583) argues that "feminist objectivity is about limited location and situated knowledge, not about transcendence." Like Harrison, Haraway ties this epistemological argument to struggles by oppressed peoples for social justice. Describing the centrality of such linkages in Black feminist theory, Collins (2000) argues that subjugated people's experiences of interlocking "axes of oppression" (e.g., race, class, and gender for African American women) can give insight into domination along other axes as well as into the broader, multi-level "matrix of domination" (248). Dialogue across such situated perspectives offers an alternative mode of objectivity, she argues, in which "partiality and not universality is the condition of being heard; individuals and groups forwarding knowledge claims without owning their position are deemed less credible than those who do" (270).

37. In this reading of Wittgenstein, I am indebted to the commentary and analysis of Floyd (2020, 2018, 2016), Moyal-Sharrock (2015), and others who have built upon Stanley Cavell's discussion of forms of life. Cavell 2015 [1969] has been especially helpful for my own thinking; see also discussion by the contributors in Martin 2018. In anthropology, Veena Das has long been in conversation with these philosophers, and her writing has also influenced my use of the concept (Das 1998). Note that my focus here is not on the specific philosophical problems that led Wittgenstein to this concept, such as the problems of nonsense or interiority. Instead, in an approach that hews close to Floyd's work, I take up forms of life as a method, or style, of analysis for elucidating puzzles that, in the abstract, seem to lead to dead ends. My aim is to show how different "*possibilities* of structuring in life" (Floyd 2020, 119, italics in original) can offer insights about key puzzles in activism and ethics.

38. This definition of ethics is taken from the philosopher Bernard Williams (1985), who makes "How should one live?" the central question of ethics, attributing it to Socrates.

39. The process of navigating these choices is what gives a life form. As I will show in the coming pages, the two broad approaches to activist life that structure environmental justice activism in Kerala are always under construction. Part of what I seek to explain is how and why they are differentiated at all.

40. These included the campaign against the Sardar Sarovar Dam on the Narmada river and the campaign to attain reparations for survivors of the spill of toxic gas at the Union Carbide pesticide plant in Bhopal, each of which has been a topic of now classic ethnographic work (Fortun 2001; Baviskar 2004).

In both cases, I was involved in student activism aimed at showing support for these movements. My role was minor, but these experiences were important to my understanding of environmental justice.

41. This slogan was the centerpiece of an award-winning Department of Tourism campaign beginning in the early 2000s (Dhanesh 2010). Its use in tourism dates earlier, as is evident from the activist and author Arundhati's Roy's reference to it in her 1997 novel, *The God of Small Things:* "So they went ahead and plugged their smelly paradise—God's Own Country they called it in their brochures—because they knew, those clever Hotel People, that smelliness, like other people's poverty, was merely a matter of getting used to." By the time I first arrived in Kerala in 2005, such ironic uses of the slogan were common among Malayalis, who often referenced it when reflecting on the virtues and vices of their own state. As Sonja Thomas (2018) notes, the slogan is also often associated with a Hindu origin myth in which Parashuram, an avatar of Vishnu, created Kerala as a paradise for high caste (Brahmin) Hindus.

42. The first site I visited when I arrived in Kerala was a protest movement against pollution and water depletion by a Coca-Cola plant in Plachimada (Aiyer 2007; Berglund and Helander 2015; Sreemahadevan Pillai 2008). This movement was one of several that inspired solidarity activism against Coca-Cola at the University of Michigan, where I would soon become a graduate student. Thus, while Kerala was new to me at the time, the linkages between home and "the field" were already dense (Appadurai 1990).

43. For example, not only the touch but even the gaze or proximity of oppressed caste bodies were considered polluting (Namboodiri 1999; Kannan 2012). Early twentieth century protest movements focused on challenging these rules by, for example, marching in prohibited roads and entering prohibited temples (Nisar and Kandasamy 2007; Aiyappan 1965; Jeffrey 1976). Harikrishnan (2023) argues that these movements were crucial to the formation of the public sphere in Kerala.

44. This includes participation in national movements like Quit India, as well as in regional protest movements like the Mappila Rebellion, in which Muslim peasants in Malabar revolted against both their Hindu landlords and their British rulers (Gangadharan 2008; Panikkar 1989).

Prior to independence, the region that became Kerala consisted of three separate territories: Malabar in the north, which was directly ruled by the British, and the two princely states of Cochin (central) and Travancore (south), which had limited sovereignty under British paramountcy. While activism for independence traversed these territorial and administrative divisions, it also took a somewhat different form in each place (Jeffrey 1993; Menon 1994). For accounts of the roles of language reform and print media in the nineteenth-century emergence of Malayali identity, see Arunima (2006) and Ambrosone (2022). For a critical account of the historicization of Malayali identity, see Devika (2008).

45. Menon (1994) shows in detail how the Communist movement in Malabar emerged from, and built upon, protest against caste discrimination and British imperialism. For an early anthropological account of Communist organizing in Kerala after independence, see Gough (1965a, 1965b, 1968b, 1968a).

46. Heller 1999.

47. This transition to parliamentary politics came with much debate, among both activists and scholars, about the implications for the Party's revolutionary agenda (Nossiter 1982). This debate was, in part, behind the eventual 1964 split of the Party into the Communist Party of India (CPI) and the Communist Party of India (Marxist), or CPM. After a period of contention, during which the CPI went into coalition with the Congress, the CPM emerged as the more powerful party and leader of the Left Democratic Front coalition beginning in 1979, with the CPI as another key member (Santha 2016). For historical and early anthropological accounts, see Devika (2007); Franke (1993); Gough (1967); Jeffrey (1993); Menon (1994); Namboodiripad (1976); and Nossiter (1982). For an account of contemporary student politics, see Lukose (2009). For discussion of the relation between environmental justice protest and contemporary Communist protests, see chapter 4.

48. Parayil 2000.

49. Lukose 2009, 28.

50. Sen 2000; Ramachandran 2000; Jeffrey 1993; Tharamangalam 2007.

51. The argument for the relevance of Kerala as a "model" originated with a study that scholars at the Center for Development Studies in Thiruvananthapuram prepared for the United Nations in 1975. In the 1990s, influential studies in in anthropology (Franke and Chasin 1992), political science (Heller 1999), and history (Jeffrey 1993) contributed to the dominance of this narrative. For critical analysis of this narrative, see Devika (2008, 2010).

52. Jeffrey 1993, 2009; Harikrishnan 2020, 2023.

53. P. Radhakrishnan 1981.

54. Heller 1999, 16. Heller argues that, by institutionalizing class struggle in its unions and party-affiliated organizations, the CPM established an ideological hegemony in the state.

55. Isaac and Franke 2002.

56. Heller 2005; Isaac 2003.

57. Lindberg 2005, 14; Devika 2010.

58. Devika 2010.

59. Rammohan 2010. But see Steur (2010); Sreekumar and Parayil (2006).

60. Rammohan 2008; Heller 1999; P. Radhakrishnan 1981.

61. See, for example, Escobar (1995); Ferguson (1994). In a separate line of criticism, some economists have long pointed to high unemployment as a drawback of the Kerala Model; they argue that widespread education has prepared too many young people for too few white-collar jobs while strong labor regulations and unionization have driven away foreign investment and led to economic stagnation (Tharamangalam 1998; Prakash 2004; Thomas 2006). While education opened up opportunities for social mobility via emigration to the Persian Gulf, Singapore, Australia, and other foreign countries, the resulting remittances have failed to fuel economic growth and employment within Kerala (Kannan 2023). The dream of development without economic growth, they suggest, was no more than a dream after all.

62. J. Devika (2007) argues that, by the late twentieth century, "developmentalism" had become an integral part of Malayali identity. Similarly, writing about social mobility among the Izhava caste, Filippo and Caroline Osella

(2000, 8) describe "a widespread ethos of mobility, now lexalized in Malayalam as *progress*" (italics in original). This ethos arguably has much deeper roots. For example, Arunima (2003) traces a long-running desire for "progress" and "modernity" in the gradual reform of matrilineal descent and inheritance among the Nayar caste going back to the mid-nineteenth century. As Arunima documents, however, this push for progress has always been haunted by disappointments and doubts, and these arguably escalated in the later part of the twentieth century. In an ethnographic study of high suicide rates in Kerala, Chua (2014, 34) argues that, by the early twenty-first century, this "dream of developmentalism has been thrown into crisis" as a growing middle class struggled to find opportunities that matched their aspirations. Thus, by the time of my own research, the widespread notion of Kerala as a place of "progress" was as much a source of anxiety as of optimism.

63. I was not alone in this. Political scientist Prerna Singh (2011, 2015) argues that widespread support for social welfare programs can be attributed, in part, to high levels of solidarity, or "we-ness," among Malayalis.

64. The term *samaram* (plural *samaraṅṅaḷ*) is commonly glossed as "struggle" among Malayalis, but this can be confusing because the English term does not necessarily have any political connotation. Other glosses such as "strike" or "movement" are also imprecise, the former because it denotes a specific tactic in English (one of many that might be used in a *samaram*), the latter because it does not necessarily capture the oppositional character of a *samaram*. For ease of reading, I will mainly rely on the term "protest" here, but "protest movement" might better capture the temporality of *samaram,* which is not just an action taken on a particular day (a march or a sit-in), but an ongoing activity of resistance.

65. This rejection of the political culture of protest is consistent with a broader transition in Kerala's political culture beginning in the late twentieth century. In an ethnography of college student politics, Ritty Lukose (2009, 140–41) has documented the rise of a neoliberal "civic public," in which some Malayalis see protest politics as an affront to the freedom of consumer citizens to participate in the market. This transition can fruitfully be understood against the background of Filippo and Caroline Osella's (2000) ethnography of changing avenues for social mobility among the Izhava caste in the 1990s. Simiarly, Chua (2014, 6) writes of Malayalis in the early twenty-first century as leading "developed lives in a developing world."

66. People's assessment of past social movements tended to follow a similar pattern. For example, those who complained about the protests of the Communist parties might speak nostalgically about the *Vimōcana Samaram* ("Liberation Protest") that opposed the early electoral success and policy initiatives of the Communists in the 1950s. In her ethnography of the Syrian Christian community, Sonja Thomas (2018) describes how this protest is seen as a landmark for a "minority rights" movement that has centered on conflict between Catholic schools and Communist education policy.

67. Chaturvedi's (2011, 2015) ethnographic research on inter-party violence in northern Kerala offers a particularly striking angle on this side of Kerala politics. She describes how workers for the Marxist CPM and the Hindu Nationalist RSS, respectively, formed tight-knit communities based on affective

ties of love (*snēham*) and practices of mutual caring. Embedded in such collectives, party workers ease their own suffering while also subsuming individual culpability for violence into the agency of the whole.

68. In practice, party identities tend to be even narrower, following the fragmentary pattern of the large number of political parties and sub-party factions in Kerala's parliamentary electoral arena.

69. This dynamic of the stick is arguably a byproduct of the drive for progress which has, in other respects, made Kerala a "model" of development. As the environmental activist and scholar Arundhati Roy (1999) has eloquently written about protest of the Sardar Sarovar dam on the Narmada river, environmental justice movements are often seen as running counter to the "greater common good" in India. The Kerala Model is classically a model that prioritizes people over economics. But by the time of my research, my interlocutors—both activists and non-activists—commonly stated that both major political party coalitions were pushing similar programs for economic growth. Activists argued that this common drive for development accounted for the failure of any major political party, despite their differences, to support environmental justice movements when in power.

70. In a study of the rise of Hindu nationalism in India, Thomas Hansen (1999, 200) writes "Within colonial epistemologies, communalism and sectarian violence were regarded as exaggerations of the 'pathologies' of the East—the uncontrollable, deeply rooted religious sentiments that made the Orient oriental. . . . This construction of communalism as the irrational force of primitive and atavistic hatred emanating from the 'masses' steeped in tradition and superstition, and easy targets for manipulators, has remained dominant within the 'educated' middle classes and the political elite in India to this day, albeit in slightly changed forms." While radical environmental activists in Kerala were unusual in making opposition to sectarianism a core purpose, many Malayalis spoke to me about the evils of communalism, priding themselves on the relative lack of it in Kerala compared to other states.

71. Francis Cody (2015, 52) argues against the tendency to view Indian public life as a "deviation, failed replication, or even crisis" from the perspective of the norms of the liberal public sphere, and recommends that scholars instead ask how "postcolonial publicity" can improve our understanding of actual politics and, even, suggest alternative normative visions of democracy.

72. Dumont 1970. Such studies have not always been clear about which side of this contrast is more desirable: the sociologist Émile Durkheim worried that a lack of community in modern life was fraying the moral fabric of society, leading, for example, to increased rates of suicide (1951), while Dumont used South Asian morality as a foil to critique the limits of Western values. Nonetheless, insofar as they contrast West and non-West, such comparative frameworks tend to introduce ideologies about moral progress into the analysis of actual moralities.

73. Fox 1989; Mines 1994; Laidlaw 1995; Alter 2000; Fortun 2001; Dave 2012. An especially powerful example is the anthropologist James Laidlaw's (1995, 2010) research on Jainism, a South Asian religion that emphasizes nonviolence toward all forms of life as a means of spiritual liberation. Laidlaw

(2010) describes how many Jains living in diaspora see their religion as anticipating the extensionist morality of some environmentalisms. He also explores key differences between Jainism and Peter Singer's environmentalist philosophy that suggest the convergences between Jain morality and environmentalism may not be as neat as they first appear. Nonetheless, the case makes clear that environmentalists have no monopoly on ideals of impartiality.

74. Anand Pandian's (2009) ethnographic study of virtue and agrarianism among a "criminal caste" in Tamil Nadu is an excellent example of how anthropologists have emphasized the importance of community while refuting dichotomies between East and West. Pandian notes that two types of community—caste and village—have classically been charged with holding back Indian moralities from advancing toward Western ideals. Taking on the question "How do people come to live as they ought to live?," he shows how modern modes of ethical self-cultivation in Tamil Nadu continue to engage and perpetuate, rather than transcend, caste and localized modes of belonging.

75. Partha Chatterjee (1993, 223) writes, "One of the fundamental elements in the colonial conceptualization of India as a 'different' society was the fixed belief that the population was a mélange of communities." On the colonial categorization and enumeration of these communities, see Cohn (1996); Dirks (2001). On the influence of colonial governance on the formation of the identities and modes of mobilization that came to constitute "communalism," see Freitag (1989); Pandey (2006).

76. Kerala has often been lauded for its relative lack of communal violence compared to other Indian states, despite relatively high religious diversity (Mannathukkaren 2016), but see Ruchi Chaturvedi's (2011, 2015) work on interparty violence for a counterpoint.

77. In this sense, this book's account of South Asian politics differs from descriptions of Indian politics as "split" between public sphere-like communicative practices and the politics of crowds, interest groups, and propaganda (Chatterjee 2004; Harriss 2011; Bayly 2009). The politics of people's protests straddles the lines that define these splits, combining print media and street politics. Nor are people's protests easily categorizable as either a politics of the subaltern or a politics of the elite. People's protests are not unique in the respect; Mitchell (2023) shows that many of the tactics of Indian street politics blur the line between what she calls the "'indoor' deliberative and associational politics" of (elite) civil society and the street politics commonly associated with the subaltern masses (36).

78. This is a loan word from Arabic, which may account for the occasional claim that, in Malayalam, it was originally used to refer to Muslims. It is primarily used to refer to white people today, echoing its common use to refer to the British during colonialism.

79. By mobility, I mean not only my ability to travel overseas, but also the gendered freedom to move about in public spaces within Kerala when I wished. For a discussion, see Ambrosone et al. (2023).

80. Henry David Thoreau's (1854) *Walden* was popular, as was Aldo Leopold's (1949) *A Sand County Almanac.* Activists also introduced me to Masanobu Fukuoka's (1975) *The One-Straw Revolution*, André Gorz's *Ecol-*

*ogy as Politics*, Tetsuko Kuroyanagi's (1981) *Totto-Chan*, and many other translated books that inspired and informed their politics.

81. Mathias 2010.

82. This is not to say that the differences between these groups were purely ideological, rather than sociological—only that the sociological differences were not easy to pin down in terms of demographics such as income level or caste status. I delve into this more deeply in chapters 1 and 2. Notably, however, this differs from some prior studies of environmental justice protest in India, especially Baviskar (2004) and Fortun (2001). Each of these studies describes similar structures of collaboration that align clearly with class distinctions—directly impacted communities from oppressed groups and "middle-class" activists working in solidarity. That this alignment was not so clearcut in this study may be due, in part, to the specific demographics of Manamur; there were other collaborations in Kerala that aligned more neatly with caste or class hierarchies. But this difference is also likely due to the leveling consequences of the "Kerala Model," especially in education.

83. Initially, I had conceived of this project as "activist anthropology" in Hale's (2006) sense, in which the researcher allies with an activist organization from the outset and seeks to produce knowledge conducive to the organization's goals. However, this was immediately rendered problematic by the highly decentralized nature of environmental justice activism in Kerala: there was no clear organization with which to ally. Moreover, the activists I met at *Kēraḷīyam*, though they presumed some degree of shared values, expressed more interest in my critical perspective as an outsider than in how my research could further any specific campaign—a position consistent with their general skepticism of group membership and emphasis on self-critique (see chapter 1). This did not mean that I felt no pressure to align my views with theirs, but it did allow for debate that might have been more difficult in the context of a more formal alliance.

84. During my longest fieldwork stint, I was investigated by several branches of the Indian police, always because (the police told me) of a complaint that I was actually organizing, rather than merely studying, the protests against the gelatin factory in Manamur. While I was ultimately able to continue my research during these investigations, this was the final nail in the coffin for any aspiration to combine activism and ethnographic fieldwork.

## CHAPTER 1

1. It is difficult to make any good estimate of how many Malayalis see themselves as belonging to this group, let alone how many might be seen so by others. For reasons that this chapter will make clear, there was no membership list. At the time, *Kēraḷīyam* had about seven hundred subscribers, but when I conducted a survey of a random sample from this group, I found that many were not very involved in other ways—though all considered themselves friends of Sunny. The key thing was that Faiza and other environmental activists saw their way of life as an alternative (*badal*) to the mainstream (*mukhyadhāra*). In this sense, they saw themselves as always in the minority.

2. Throughout the text, references to the Communist Party and Communists denote the CPM when describing contemporary events, unless another Communist party is specified. As noted in the introduction, the rise of the Communist Party of India (CPI) in Kerala was crucial to shaping the state's political culture. When the CPI emerged as the ruling party in Kerala's first elections in 1957, it faced an identity crisis over its revolutionary agenda vis-à-vis its new power as one party in a representative democracy. This debate led to a split into the CPI and the Communist Party of India (Marxist), or CPM (Nossiter 1982). The CPM ultimately emerged as the larger political force, leading the Left Democratic Front in Kerala's coalition system. For an account of contemporary student politics, see Lukose (2009).

3. Nonetheless, his voice took a wistful tone (perhaps mere nostalgia, but maybe also a little regret) whenever he talked about the magazine—especially when he recalled how, in 1992, he had asked his father why *Sovietland* had stopped coming to their house. His father had told him it would never come again, but he had not wanted to explain why. It was only years later that Adarsh finally understood.

4. Vegetarianism was common among this group, but could also be controversial because vegetarianism was historically associated with upper-caste dietary ethics and could be seen as an assertion of caste superiority (Srinivas 1966; Klein 2008). Much discursive work was done, therefore, to disconnect environmentalist vegetarianism from any Brahminical overtones. For more discussion of environmentalist dietary ethics and caste distinction, see chapter 3.

5. When I talked with people outside these activist circles, they expressed disbelief that such disavowals of knowing caste status could be sincere. Some suggested tips for inferring caste from physical features, dietary preferences, or household furnishings. Nonetheless, even after a year of living with Adarsh, he consistently denied knowing the caste of most other activists. An exception were activists in Dalit or tribal movements, for whom being open about caste identity was part of their activism. Adarsh argued that, even in other social settings, many Malayalis are overconfident in their ability to infer caste. When, in the pursuit of rigorous sociological description, I tried to apply the advice of others and make my own inferences, he laughed at the hopelessness of my attempts. In activist circles, where caste markers were carefully avoided, he believed the only way to know was to ask directly. And this he did not do.

6. Kerala's robust government-provided basic education and social welfare programs, discussed in the introduction, produced the conditions for the possibility of both these life paths. Francis and Saleem both completed school through the tenth grade, the level through which education in Kerala is free and mandatory.

7. Thus, as Faiza suggested in the book's introduction, choosing to be an activist could also be a departure from life paths that prioritized certain kinds of jobs and strategies for financial security and social mobility. But this could also entail other job opportunities and paths to security. Similarly, Morton (forthcoming) argues that when Brazilian peasants turn away from "growth" and leave high-wage jobs in the cities to pursue agrarian livelihood in rural areas, they are not necessarily rejecting financial wellbeing as such. Rather, he

holds, they have a different vision for financial wellbeing that emphasizes freedom and permanence.

8. Compare this with the stories of departure and activist becoming through which the history of Kerala's Communists has often been narrated—primarily stories of high-caste Namboodiri and Nayar Hindus (e.g., Namboodiripad 1976). Jeffrey (1978) argues that legal reforms dismantling matrilineal norms among Nayars led to a generation of Nayar men who were cut off from their communities, priming them for Communist ideology and activism. Menon (1994, 132) agrees on the importance of these high-caste young men to the early Communist movement, but argues their activism was not a response to deracination but an effort "to bolster their vestigial status by acting in conjunction with their erstwhile dependents."

9. In keeping with their skepticism of party politics, many of the activists described here did not vote. However, if and when they did vote, they generally reported voting for the CPM or other LDF parties. This changed, however, with the emergence of the Aam Aadmi Party in the 2014 national elections. The Aam Aadmi party emphasized solidarity with people's protests, and Medha Patkar and other leaders in the National Alliance of People's Movements (NAPM) ran on the Aam Aadmi ticket in their own states. Still, beyond those few activists most connected with NAPM in Kerala, most of those I spoke with were skeptical of these alliances and did not participate in the 2014 elections.

10. This use of the term *badal* to describe people's politics signified, most directly, a rejection of organized parties and interest groups. But it also encoded other notions of alternativeness with regard to economic behavior, the management of one's body, relations with family and friends, and other aspects of life—meanings I examine more closely in chapter 3.

11. In keeping with usage within Kerala, I use the capitalized "Leftist" to denote the politics of Kerala's Communist parties and the Left Democratic Front. I use the lowercase "leftist" in the broader sense including ideologies of liberalism, progressivism, socialism, and lower-case communism—without any necessary connection to a party.

12. Panangad 2018. This movement emerged against the background of earlier uses of print media by caste reform and Communist movements. Whereas the Communists sought to extend their reach to the very limits of the reading public and beyond—especially through the use of reading rooms (Menon 1994; Harikrishnan 2020)—the little magazines fostered more limited and relatively elite publics. Nonetheless, relatively high rates of literacy in Kerala (which afforded and were bolstered by earlier uses of reading in social movements) also meant that a large proportion of the Malayali population could, in principle, participate in little magazine publics. Indeed, by the period of my research, some observers argued that more widely circulating weekly magazines had supplanted little magazines as bearers of avant garde literature, cultural critique, and political analysis (e.g., Shrijan 2012).

13. Little magazine publishers positioned their magazines as an alternative to the influence of party discipline in Left publications. Tensions between artistic freedom and ideological uniformity had been a topic of controversy among

Leftist writers and intellectuals since the early years of the Communist movement (S.P. 2012).

As renowned little magazine editor Civic Chandran (2012) notes, the term "little magazine" (an English loanword used to describe this genre of small-circulation periodical) is something of a misnomer. In their heyday, these magazines published many of the biggest ideas and most influential literature by the most renowned thinkers and artists in Kerala. What made them "little" were the small communities of intellectuals that they helped to sustain—including the Vanchi Lodge community described here, in which Chandran had a key role.

14. M. Roy 1952; M. Roy and Philip 1968.

15. George 2008.

16. Radhakrishnan 2008.

17. For example, Govindan (2008a [1978]).

18. These magazines included *Prēraṇa, Vākk, and Pāṭhabhēdam.* In addition to Vanchi Lodge itself, another key thread was Civic Chandran, who served as editor of these magazines, and whose own ideological shifts paralleled (and arguably influenced) those of Thrissur-based activists. The building containing Vanchi Lodge was condemned and bulldozed in 1992, marking the end of the first iteration of *Pāṭhabhēdam.* When Chandran began publishing *Pāṭhabhēdam* again in 2002, now out of the northern Kerala city of Kozhikode, he envisioned it more as a forum for literature and cultural criticism than as a catalyst for activist mobilization. This opened the way for *Kēraḷīyam* to be the primary social hub for activists in people's protests, especially in central Kerala.

19. *Prēraṇa* was the organ of the *Janakīya Samskārika Vēdi,* or People's Cultural Forum (PCF). From 1978 to 1982, the PCF became a hub for alternative leftism in Kerala, and drew much larger and more ideologically diverse participation than the Maoist party to which it was officially attached. This was arguably the point at which the activist lineage I trace here was most unified and robustly institutionalized, though this unity did not last long. Retrospectively, activists I interviewed attributed the PCF's 1982 dissolution to conflicts with the party over ideological discipline. For a concise history and analysis, see Sreejith (2005).

20. Sreejith 2005.

21. One reference point for this genealogy is the *Save Silent Valley* movement, which sought to prevent a hydroelectric project that would have flooded a large expanse of forest (Gadgil and Guha 1994; Manjusha 2016). Some activists called this Kerala's first people's protest, though this was debated.

22. Jacob 2010.

23. A professor of zoology, Jacob founded a zoology club for college students in 1972, which became SEEK in 1977 (Shaji 2020). In 1980, SEEK began publication of *Sūcimukhi,* commonly known as Kerala's first environmental magazine, which remains in print today.

24. A catalyst for the convergence of the post-Marxist and conservationist streams was the emergence of widely-publicized environmental justice protests elsewhere in India—especially the protest of the Sardar Sarovar Dam on the Narmada river (Baviskar 2004) and the campaign on behalf of survivors of the 1984 Bhopal gas spill (Fortun 2001). Activists in Kerala were inspired by these movements, developments in which were highly publicized in the mid-1990s,

when Sunny and others founded *Kēraḷīyam*. The framing of the Narmada protests as a "people's movement" and the eventual formation of the National Alliance of People's Movements (NAPM) coalition also influenced the discourse and practice of "people's protest" described here.

Nonetheless, at the time of my research, Kerala's people's protests were, in many ways, disconnected from the national discourse associated with NAPM, which had had little success in bringing them into its coalition. Part of the disconnect was linguistic. When NAPM held its national convention in Kerala in 2012, national organizers were frustrated by a lack of local participation, which they attributed to a lack of interest. While there may have been some truth to this, Malayali activists also described to me the difficulty of communicating with these predominantly northern visitors, who used Hindi as a lingua franca. In addition, Kerala's people's protests are usually directed at state, district, or even *pañcāyatt*-level politicians and government officials. Thus, they rarely share the same targets with campaigns outside the state. Finally, as suggested above, while activists in Kerala's people's protests work on similar issues to activists in NAPM organizations, they draw on social movement genealogies that are in many ways unique to Kerala. Consistent with the post-Marxist tradition described here, Kerala activists were highly skeptical of any attempt to form robust institutions or membership groups, and many applied the same skepticism to participation in NAPM.

25. This process can be thought of as "framing,", as theorized by Goffman (1974) and adapted by late-twentieth-century sociologists studying social movements (Benford and Snow 2000; Snow and Benford 1988). In keeping with the latter literature, the "making" of people's protest is a strategic meaning-making process aimed at advancing activists' aims. However, the framing process also reshapes those aims; solidarity with people's protests becomes a central cause of Kerala's environmental activists (Polletta 1997, 2008). In this sense, the literal looping of discourse between the front and back of the *Kēraḷīyam* office generates "looping effects" in Hacking's (1999) sense. The representation of reality also contributes to producing the reality that these activists (as well as others—see chapters 2 and 4) inhabit.

26. This intermingling of print media and discussion parallels some of the communicative norms described by Jurgen Habermas in the *Structural Transformation of the Public Sphere,* which he argues were foundational to the emergence of modern ideals for liberal democracy in Europe. These parallels are not coincidental—M. Govindan and his interlocutors described their magazine-based discourse as contributing to a "renaissance" (*navōtthānam*) in Malayalam literature and politics that explicitly built on historical narratives of the emergence of modernity in the West (Govindan 2011c; see especially the essay on poetry and renaissance, Govindan 2011d [1974]). For critical discussion of the notion of a Malayali renaissance, see Devika 2008). Both Govindan and the activists described in this book were also highly critical of many aspects of modernity. Nonetheless, when taken together with the environmental activists' project of extending the boundaries of moral concern, one might see *Kēraḷīyam* as heir to a modernist project—a case study, even, in the possibilities and limitations of Western ideologies about moral progress. Such a reading would likely prove stultifying, however, because—inasmuch as this is an Indian case—any shortcomings of

these activists' projects could be referred back to differences between India and Europe, perpetuating (post)colonial depictions of India "in terms of its essential lack" (S. Roy 2007, 107) rather than stimulating thought about what may be lacking in the theory of Habermas. Instead of measuring Kerala activists against Western ideals, then, I focus on understanding the conflicts that arise within activists' own lives and on their own terms. This opens the way to asking how their lives may speak to the lives of others without reducing such comparison to the question of whether, or in what ways, they are "modern."

27. Govindan 2008b [1948].

28. Not all activists in the *Kēraḷīyam* scene necessarily saw their discourse about "the people" as having Maoist roots. Notably, however, one activist who had been central to the Vanchi Lodge tradition but was not connected with *Kēraḷīyam* criticized the use of the "people's protest" label for environmental conflicts because he regarded this label as a definitively Maoist term. He preferred to call the environmental justice protests "new social movements," but most of those on the *Kēraḷīyam* scene were less familiar with this usage (cf. Sreekumar and Parayil 2006, 2010).

29. This communicative theory shares much with the activist pedagogy of Freire (1970, 1974), who made dialogue a mechanism for critical consciousness of oppressive social conditions. While Freire was not much discussed among activists connected with *Kēraḷīyam,* his ideas were important to literacy movements in the region. In an ethnography of literacy activism in Tamil Nadu, the anthropologist Francis Cody (2013) describes how activists struggled to use literacy to "raise consciousness" without simply disciplining students to conform to activists' own notions of consciousness. These tensions are already present in Freire's own work, particularly where he delves into the practical details of implementing his pedagogical ideals (1974).

30. This parallels arguments within the US environmental justice movement (against the claims of some mainstream environmentalists to have a more expansive moral outlook) that environmental justice is broader, or more "inclusive," because it extends environmentalism to issues and populations overlooked by predominantly white mainstream activists (Taylor 1993).

31. Chowdhury 2019, 8.

32. Deleuze 2019.

33. Laclau 2005; Morgan 1989.

34. Gayatri Spivak (2003) famously pointed to a similar puzzle at the heart of the Subaltern Studies historical project, which sought to trace the history of the political agency of the people during British colonialism (Guha 1982). Writing the history of the people necessarily meant conforming the heterogeneity of actual people to the ideal of the whole, but doing so inevitably produces new misrepresentations and exclusions.

In this book, I do not attempt to study the people as such, but instead give an account of how some people work to constitute themselves or others as the people. In this, I build on the work of Cody (2015) and Chowdhury (2019) on popular politics , who call for greater attention to reflexive production of the people as a way of moving beyond preoccupations with Western normative ideals in the study of South Asian politics. In this chapter, I describe the constitution of the people as an object of solidarity via small-circulation print media. In chapter 4, I describe a complementary process, in which some people use a range of tactics, in person and mass-mediated, to produce the people as an object of sympathy.

35. *Janaṅṅaḷ* is the plural form of *janam*, which is also the root for *janakīya*, "of the people," in *janakīya samaram*, "people's protest." Unlike *āḷ*, "person" (plural, *āḷukaḷ*, "people"), the singular *janam* and the plural *janaṅṅaḷ* both mean the people as a collective or mass. In Communist literature, the related term "*bahujanaṅṅaḷ*," or "masses," is used in a similar manner—sometimes intermingled with uses of *janaṅṅaḷ*—where the prefix *bahu-* indicates "numerous" or "plentiful" *janaṅṅaḷ*. Adarsh said that it would be incorrect to use *bahujanaṅṅaḷ* to describe contemporary people's protests, which mobilize relatively small numbers of people. These are "the people" but not "the masses."

36. Laclau 2005.

37. Environmental justice activists—both radicals and locals—made similar contrasts between "the people" and elites or powerholders. Canovan (2005) attributes this notion of "the people" to populism, but solidarity activists' ideal of the politics of people's protest differed from populism in two key ways. First, this genre of people's protest was not a mode of mass politics. Second, solidarity activists distinguished sharply between "the people" and the "common people" (*sādhāraṇakkār*).

38. Partha Chatterjee (2020) notes that such semiotic emptiness (what I call thinning) is crucial to bringing distinct demands by diverse groups under the rubric of "the people." Chatterjee argues that "the people" and "the enemy" can then become floating signifiers, retaining the rhetorical power of the distinction but refilling it "with fresh content in order to stay relevant to the most perceptible dividing line separating the haves from the have-nots" (144). Yet Chatterjee attributes this process to "political society," a domain of Indian politics that he distinguishes from the liberal democratic norms of "civil society." As noted earlier, however, the uses of small magazines by Kerala activists seem to align with many of these norms, especially as theorized in the work of Habermas. This politics of "the people" does not seem to slot neatly into either side of Chatterjee's civil/political distinction.

39. These reading practices have strong family resemblances with the Gandhian theory of "slow reading" described by Isabel Hofmeyr (2013) in her history of Gandhi's work with the periodical *Indian Opinion*. Like many of the other traces of Gandhian praxis in environmental justice activism, Kerala activists did not necessarily attribute their notions about reading and *anwēṣaṇam* to Gandhi and his followers. But they did see their reading practices, like their publishing work more generally, as part of a tradition that reached back to the earliest uses of print media in social movements, including the nationalist movement and the Communist movement.

40. Among commonly read materials were books and pamphlets criticizing traditional schools. Those activist parents who withheld their children from school were viewed with admiration by other activists, who marveled at how much happier and more learned these children were than their peers (see chapter 3).

41. Likewise, Dhanya also described her people's protest activism in the idiom of inquiry, comparing it to my research. "Just like you're inquiring, I'm

also inquiring," she said. "It's just not with an academic lens. . . . The people's problems are brutal; day-by-day, they continue to be brutal. We must try to discover solutions. That's the inquiry I'm conducting."

42. This process is well-described by Carl Boggs's (1977) notion of "prefigurative politics," in which activists attempt to bring about social change by producing in their relations with one another the forms of sociality that they would like to bring about in society at large. The term has since been used to describe modes of activism that put strong emphasis on organizing process, even making process the primary end of organizing (Breines 1989; Polletta 2012). Prefigurative politics have been important to the global justice and Occupy movements (Graeber 2002, 2009; Yates 2015). In Kerala, radical environmental activists used "alternative" practices to prefigure not only social relations, but also ecological relations (see chapter 3).

43. Environmental activists were far from the only people using the term *kūṭṭāyma*. Indeed, one Malayalam professor complained to me that the term had become an annoying trend. The gloss of *kūṭṭāyma* as "fellowship" appears to have deeper roots in Christian discourse. Compare also the gatherings for reading and discussion Menon and Harikrishnan (2023) describe in the northern Kerala city of Kozhikode in the 1960s, which they also term as "*koottaymas*—gatherings of informal nature." Recently, the term has become common in the titles of neighborhood organizations and volunteer groups as well, including some contemporary caste-based organizations.

44. In some sense, these activists aspired to an ideal that Dave (2015), writing about animal activists in Delhi, has called "indifference to difference" (see also Davé 2023). But much depended on which differences were treated with indifference. As Sunny's speech indicates, they also very much valued being different—a value evident in their criticism of the mainstream and ordinary people. Their indifference to some differences (of caste, religion, party) itself became the difference that they valued above all.

45. Compare Gatt's (2017, 141) description of ideals about unity among members of Friends of the Earth, a global organization of environmental activists. There as here, "varying, possibly conflicting ideas about environmentalism [are] valued," while unity and belonging are found in "fighting for a cause." In many ways, FOE activists' efforts for ethical self-formation parallel the practices of activist inquiry and community-building described here, though in a more institutionalized context.

## CHAPTER 2

1. *Nāṭ* is part of a broad grammatical category that linguistic anthropologists call "shifters," parts of speech that change meaning depending on various aspects of the context of use, such as the speaker, the addressee, or the place in which speech occurs (Silverstein 1976). More specifically, the term belongs to a subcategory of shifters that encode place-based belonging (Stasch 2009, 45, 46).

2. The most common Malayalam term was *maṇam,* a word that, like the English "smell," does not carry a strong positive or negative connotation. Peo-

ple also sometimes said *durgandham* and *nāṯṯam*, both of which are always negative (like the English "stench") as well as the English "smell."

3. Douglas 1966; McKee 2015.

4. Thus, "local" protest, even as it responded to experiences of living with pollution in a particular place, was always an interaction with "global" actors, even before the arrival of solidarity activists (Tsing 2000). To be local was to engage in these interactions in a local way, not to be separate from the global (Lambek 2011).

5. Nayars are a *savarna* caste (that is, relatively high status) and one of the most populous caste groups in Kerala.

6. Technically, the Manamur factory produced ossein, a fibrous protein that is a major component of bones, which was taken to a second factory for making gelatin. To extract the ossein, the bones were baked in ovens and then soaked in hydrochloric acid until the mineral content dissolved. The remaining liquid was then filtered and distilled. This produced a great quantity of gaseous emissions released from smoke stacks visible high over the factory walls. Another byproduct was a blackish sludge having the same smell, which was emitted via an underground pipe into a nearby river.

7. Freire 1974, 1970; MacKinnon 1982.

8. The usage I gloss as "raising consciousness" used two transitive verbal clauses: *bōdhavalkarikkuka* and *bōdhyappeṭuttuka*. Early in my work, I was so accustomed to hearing solidarity activists use these phrases similarly that—despite finding somewhat different translations in my Malayalam-English dictionary—I had difficulty understanding the differences between them. When I asked native speakers, some also struggled to explain the difference to me. Later, however, I found that Sunitha and other Manamur protestors did not mix these terms; in returning to my recordings, I found that they rarely spoke of *bōdham* or *bōdhavalkarikkuka*, though they often spoke of *bōdhyam* and *bōdhyappeṭuka*.

9. Shapiro (2015) describes a similar process of attunement to chemical exposure in the United States; see also Ahmann (2018).

10. In an ethnography of experiences of bodily pain in Micronesia, anthropologist C. Jason Throop (2010) argues for the centrality of attention to "the transformation of painful sensations into meaningful, morally valenced, lived experiences" (10). Throop shows how culturally specific modes of attending can, thus, begin the process of transforming "mere-suffering" into "suffering-for," in which enduring pain can become a virtuous practice of moral responsibility and compassion (275). A similar process can also be tracked in Manamur protestors' stories of how experiences of pollution led them to activism. In addition, however, the influence of their activism on their attunement to pollution suggests that the directionality of this process can be reversed. At least in some cases, practices of "suffering-for" can reinforce, direct, or intensify attention to suffering.

11. Ahmann (2020) points out that "'the community' of the impacted is a working category," the definition of which is central to the process of environmental justice activism. In an ethnography of activism against an incinerator in Baltimore, she describes how different definitions of the impacted enabled different strategies for challenging the incinerator.

12. Lambek 2011.

13. At the time of this research, the religious makeup of the administrative subdistrict in which the gelatin factory was located was approximately 57 percent Hindu, 35 percent Christian, and 8 percent Muslim (Census of India 2011). There was no official membership list for the Manamur Action Council, but participation roughly reflected the broader demographics of the area. Among Christians, Syro-Malabar Catholics predominated in the area, and this was reflected in the Action Council. Among Hindus, participation likewise seemed to approximate demographic patterns for the area. There were a large number of participants from the Ezhava caste and the Vishwakarma set of castes (especially traditional carpenters and clay workers), with a minority of participants from the higher-caste Vermas and Nayars as well as the oppressed caste (Dalit) Pulayas.

14. In describing such bias, Sujit and other oppressed caste participants in the Action Council did not often point to specific, concrete acts of discrimination, and it was not clear whether this was due to hesitance to pinpoint people or to the subtlety of casteism. Writing about cross-caste relations among neighbors in eastern Kerala, Thiranagama (2019) notes that even though higher-caste Izhava and oppressed-caste Dalit women formed friendships while conducting agricultural work together, the Dalit women were highly aware of casteist behavior among their counterparts. She writes: "Their caste-sharpened eyes see minute movements within the texture of everyday life, things hard to put one's finger on precisely, but ever-present and constitutive of how one thinks about others."

15. Women in Kerala are well known, both among development scholars and the broader Indian public, for their high levels of literacy and political participation (Jeffrey 1993). Nonetheless, social movement activism, running for elected office, and other political activity has often not entailed transformation of patriarchal norms in everyday life, such as clothing norms or roles in housekeeping and childcare (Devika 2007; Thomas 2018). Within environmental justice campaigns, specifically, local women are often put forward as figureheads while men control decision making processes (Binoy 2014; Devika 2010).

16. One might argue, on the other hand, that the radical environmental activists would have eventually moved on anyway. Writing about a similar collaboration in the movement against the Sardar Sarovar Dam in northern India, the sociologist Amita Baviskar (2004, 280) writes, "No one comes through anymore; there aren't any meetings. Villagers *miss* the activists, their spirited energy and their counsel, the news and connections they brought to the rest of the world. The tide of activity has ebbed. People feel stranded."

17. Williams 1985. Along the same lines, some anthropologists have viewed the study of ethics as a correction for a longstanding overemphasis on interest theory in the field (Lambek 2010; Ortner 1984, 2016).

18. Sessions 1995.

19. Research on localized opposition to infrastructure—both renewable and otherwise—has consistently found that NIMBYism poorly describes the motives and ideologies of local activists (Burningham 2000; Devine-Wright 2011; Eranti 2017). Dowie (1996) describes how the NIMBY acronym was invented by a waste incineration industry executive to discredit grassroots environmental opposition. Nonetheless, the notion of NIMBY lingers around environmental

activism, perhaps in part because of the difficulty of distinguishing NIMBYism from locally-defined environmental justice activism like that of the Manamur Action Council (Melosi 2000).

20. My discussion here draws most heavily on Joan Tronto's (1993) conceptualization of an ethic of care, which critiques and builds on Carol Gilligan's (1993) original formulation. Tronto's theory seeks to move discussion of a care ethic beyond the question of whether there is a specific "women's morality" to explore instead how politics and morality, generally, can include "the values traditionally associated with women" (1993, 3). Relatedly, while Tronto's work has been taken up by some anthropologists studying the ethics of medical care (for example, Kleinman 2009; Mol 2008; for a review of related work, see Buch 2015), the ethic of care is not specific to the set of human activities typically called caregiving or care work. Such activities may be conducted with an ethic of care, but they often involve other moral paradigms as well. Likewise, here I see care as an approach to ethics that, while more characteristic of some lives than others, nonetheless speaks to aspects of social life that are likely experienced byall humans.

21. Such extension is already anticipated by Tronto's (1993, 103) theory, which includes non-humans in the "complex life-sustaining web" maintained by care. María Puig de la Bellacasa (2017) extends this conceptualization by exploring how care traverses human and nonhuman relations, including how nonhumans care for both human and nonhuman needs (see also Ticktin 2019). Whyte and Cuomo (2016) describe how such extension of care to water, land, and other nonhuman entities is theorized within indigenous environmental movements, which they argue lay the groundwork for the integration of feminist care ethics and environmental ethics.

Such extension may sometimes begin from reconsidering what counts as relational distance. In an ethnography of human-animal relations in mountain villages in northern India, Radhika Govindrajan (2018) describes how care traverses species distinctions via relations of intimacy, attachment, and love. Govindrajan's ethnography poignantly extends what anthropological studies of kinship have long shown: to say that care is calibrated to degrees of relatedness is not necessarily to privilege caring for a specific set of people or beings, such as one's own children—because "relatedness" is itself open to question. As we see in the contrasting stories of Rabindranath and Stevenson, people can measure social distance in many ways, and they may situationally change what counts as the right kind or degree of relatedness.

22. Tsing 2005, 58. Such notions of scale have long been important to anthropological studies of the environment and environmental politics (see, for example, Tsing 2001; Choy 2005). While every point of view presupposes some scale, scales are not simply ready and waiting for us in the world—they must be continually made and remade (Carr and Lempert 2016). Chapters 1 and 2 each describe the processes by which activists produced, inhabited, and sustained specific scales of environmental justice.

23. Here, I use the term scale somewhat differently than it has commonly been employed in the literature on environmental politics. In that literature, scales are often synonymous with spatial levels or geographic ranges—for example, a par-

ticular actor operates primarily at a local, regional, or global scale. The Manamur Action Council's person-centric scaling of their campaign points to the limits of this conceptualization of scale, showing how actors at different "levels" may also be working with different scalar configurations—different perspectives on the relation between local and global, little and big, narrow and broad.

24. Ingold 2000.

25. Ingold (2000) calls these two modes of engagement with the world "building" and "dwelling," a distinction he takes from Heidegger (1971). In building, humans engage reality as "an external world of nature" and work upon this object, constructing and inhabiting "houses of culture" (Ingold 2000, 178, 179). In dwelling, humans engage the world from within the world, from a position of "enmeshment within an all-encompassing field of relations" (187). Like Heidegger, Ingold also takes dwelling to be basic to the human condition, adopting the "sphere" perspective as the perspective from which his own analysis begins—and, thus, at times seeming to present "global" perspectives as artificial or mistaken. Here, by contrast, I do not decide between these perspectives but consider how each offers a critical angle on the other.

26. Elsewhere, I have critically examined this assertion within the field of environmental social work (Mathias, forthcoming). More broadly, it is part of the history of marginalization of environmental justice movements in the US (see introduction).

27. Keane 2016.

28. To be clear, these effects result from one's perspective—the detached position from which one gazes down at the world—rather than the actual breadth of what one takes in. Compare, for example, the map of Kerala's people's protests (map 2) to the EJAtlas, a map constructed by European activists of every environmental justice protest in the world (Temper, Del Bene, and Martinez-Alier 2015). The latter is far "broader" in the sense of how much the map takes in. But in comparison with the onion-like perspective of "fighting for the *nāṭ*" the Kerala map is no less detached, no less flattening of differences. Here, "global" and "broad" indicate not how much one takes in—it is possible, after all, to look out at everything from the center of one's onion—but how one distributes value across what one takes in.

29. Anthropologists have arguably helped to lay the groundwork for these temporalizations of moralities by arguing that human ethical values have, over time, tended toward more global views of ethics. In an early anthropological study of reciprocity, Marshall Sahlins (2017 [1972], 181) argued that the person-centric organization of moral obligations, while present in all societies, "seem[s] to prevail in primitive communities" relative to "our own" society, in which norms tend to be more "absolute and universal." Perhaps out of concern to distance his point from narratives of moral progress from primitive to modern, Sahlins also notes that "the contrast with the absolute standards of the Judeo-Christian tradition is probably overdrawn: no moral system is exclusively absolute (especially in wartime) and none perhaps is entirely relative and contextual" (181). Yet he suggests that ethnologists have widely noted this contrast between their own moralities and the "sectoral" organization of moral and economic relations among those they study, for which he offers his own diagram of concentric circles for house, lineage, village, tribal, and intertribal groups.

More recently, Keane (2016) has described how efforts to "rationalize ethics under an organizing principle"—taking what he calls a "God's-eye view"—have been central to significant historical developments, including the spread of scriptural monotheism and communism (210, 239–40). Likewise, Ingold (1993) situates his account of the emergence of "the environment" within a broader analysis of the increasing dominance of globe-like perspectives in Western society.

None of these anthropologists suggest that such changes are desirable. Nonetheless, such histories provide a background to the sense of a linear trajectory, a one-way road, which has been taken up by some environmentalists as a story of progress.

30. *Næss* 2008. See also Drengson 2008; Fox 1995.

31. Carr and Fisher (2016) argue that such interscalar linkages (here, between a spatial scale and a chronological scale) can contribute to making certain scalar perspectives compelling.

32. McGurty 2009, 79.

33. McGurty 2009, 83.

34. McGurty 2009, 135.

35. As Haraway (1988) reminds us, all perspectives have their limits, all views are views from somewhere. One limit of the solidarity activists' view might be seen in their focus on Kerala, suggested by the boundaries of their map of people's protests. Such a focus could seem parochial from the perspective of activists in Delhi, for example, who might be working with maps of India, South Asia, or even the world (see Temper, Del Bene, and Martinez-Alier 2015). A "national" activist once visited Faiza and Adarsh and seemed to see their activism in this way. But the Manamur Action Council members were concerned with a different limit of solidarity activists' perspective. For them, the very project of making maps of people's protests (and the environmentalist magazines that hold them) positioned solidarity activists as a particular set of people with a particular way of looking at things and a particular agenda. In this sense, the "environmental activists" were no different than any other group that offered assistance in the campaign against the gelatin factory.

36. For example, after reading a draft of this chapter, Adarsh stated that he dislikes talk of "raising consciousness" (*bōdhavalkarikkuka*) and claimed that this phrase cannot be found in any of his or Sunny's own writing in *Kēraḷīyam*.

37. Such constraints are inherent to the politics of environmental protest. For example, working with Ingold's Heideggerian conceptualization of "dwelling," McKee (2015) argues that dwelling is inherently political. Writing about the struggle of Bedouin communities in Israel to contend with pollution and, simultaneously, to counter racialized discourse about their community as inherently polluted and disordered, she shows how Bedouins' modes of inhabiting, reshaping, and talking about their surrounding landscape were constrained by public discourse and state land-use policies.

38. Such attempts to depict protestors as outside agitators are, of course, one of the most common strategies of those targeted by protest. A classic example is the reaction of white church leaders in the US South to Black protest for civil rights—in response to which Martin Luther King, Jr. (1997 [1963]) wrote, "Injustice anywhere is a threat to justice everywhere."

39. Purposes were, thus, performances in Goffman's (1959) sense of presentations of self that strategically anticipate effects on specific audiences, rather than merely transparent reflections of the inner feelings or intentions of protestors. But I do not mean to suggest a clear boundary between inner and outer, front stage and back, such that protestors' statements about their motives should be taken as insincere. Rather, here and elsewhere, I treat the performance of purpose as partly constitutive of purpose. As Polletta (2008) argues, the frames which activists strategically construct to further their ends may also transform those ends.

40. In this way, this case is different from the moral crisis anthropologist Joel Robbins (2004) describes among Christian converts in Papua New Guinea. Robbins examines how the Urapmin struggled to reconcile their indigenous religion, which gives paramount importance to commitments to kin and close relations, with a form of Protestant Christianity concerned primarily with individual sinfulness and conformity to the will of God. The Manamur case likewise hinges on a conflict between person-centric, relational ethics and a "God's-eye-view" morality. But this is not a conflict between two moral systems brought into historical convergence. Rather, it is a conflict within one discourse of environmental justice, in which each side requires yet refutes the other.

41. The term *adivasis*, coined in the early twentieth century from the Sanskritic roots *adi* (first) and *vasi* (inhabitant), was used interchangeably with "tribals" to denote indigenous communities in Kerala (Baviskar 1997). For critical discussion of categories of indigeneity in India, including their importance to environmental politics, see essays in Karlsson and Subba 2006. On the emergence of *adivasi* identity and the politics of indigeneity in Kerala, see Steur 2011, 2017.

42. Studying collaboration in environmental movements between urban, middle-class activists and tribal peasants in the late 1990s, Baviskar (1997, 222) similarly describes how some tribal leaders (especially those more "well-off") began "questioning the authority of the activists to represent others and are modifying the very terms of the debate by linking it to issues such as dalit identity."

## CHAPTER 3

1. While South Indian food is sometimes stereotypically thought of as vegetarian, Kerala cuisine commonly includes meat. Even beef-eating, though more common among Christians and Muslims, was practiced by most Hindus I met as well. Nonetheless, the pan-Indian association of vegetarianism with Hinduism (and, especially, with Brahminism) was still evident in Kerala. If one meets a Hindu who refrains from eating meat, this could reasonably be taken as an index of piety and, depending on what other markers are evident, as a performance of high caste status. This was a source of tension among Kerala's environmental activists. Even as diet was often a focus of their activism, many were vociferously opposed to any seemingly "Brahminical" notions of morally superior diet, especially any special emphasis on avoiding beef. For more on the nuanced cultural politics of food in Kerala, see Osella and Osella (2008).

2. See the introduction for a discussion of the term *sādhāraṇakkār*, which I gloss throughout as "common people" or, occasionally, "ordinary people." Here, it is worth noting the tension between two distinct uses of this term, both of which marked a contrast with "activists" (*pravarttakar*). First, *sādhāraṇakkār* could be approximately synonymous with *janaṅṅaḷ*, "the people," who were understood to be oppressed by corporations and the state, and on whose behalf the radical environmental activists took themselves to be working. This usage was more common among nonactivists and the local activists in Manamur than among the radicals. Second, *sādhāraṇakkār* could also mean those who lived the "ordinary" life that the radicals saw as perpetuating social inequality and environmental destruction. In this usage, *sādhāraṇakkār* were not elites, but they were nonetheless distinguished from the people (see chapter 2). At times, the term seemed to be interchangeable with "middle class." Note how the contrast between these two meanings of *sādhāraṇakkār* mirrors the tension described in this chapter between working among others and setting oneself apart.

3. In his influential analysis of social stigma, the sociologist Erving Goffman (1963, 5) coined the term "normals" to denote those who do not bear the "undesired differentness" of stigma. The social predicament of the activists described here troubles Goffman's distinction between desired and undesired differentness. Radical environmental activists worked to make themselves different and valued their social distance from "common people" (*sādhāraṇakkār*). Nonetheless, that distance could be painful and, potentially, counterproductive to activists' own aims.

4. In Kerala, the term "English medicine" refers broadly to those forms of medicine associated with the global pharmaceutical industry, elsewhere called "modern," "Western," or "allopathic" medicine. In Kerala, homeopathic and ayurvedic medicine were the major rivals to English medicine. While English medicine dominated the industry, these and other healing practices were widely available and regularly used, most often in combination with English medicine. At the time of this story, Faiza and Adarsh had not taken Tara to any professional doctor, whether allopathic, ayurvedic, or homeopathic.

5. Recent anthropological studies of ethics have stressed that even "ordinary" or "everyday" social life is thick with ethical import (Das 2010, 2012; Lambek 2010). This book gives attention to the ordinary insofar as this is understood as a mode of analysis that approaches philosophical questions by situating them within forms of life (that is, in Wittgenstein's (1953) sense; see introduction). In addition, my analysis of ethics is partly in line with calls for attention to the ordinary as an empirical topic insofar as I explore how the lives of activists (and, in some cases, nonactivists—see the conclusion) are replete with ethical stakes. In the framing Faiza gives us in the introduction, one must continually choose to live for environmental justice. Yet activists also see their lives as unusual, out of the ordinary, or even extraordinary; and so they are seen by others as well. Thus, the mundane ethical concerns I describe in this chapter—air conditioning, Coca-Cola, tea, VapoRub—actually speak to a different point. Radical environmental activists in Kerala seem obsessed with "ordinary" things, but this is what makes them alternative, weird, and uncommon. They inject

ethical import into things that are *not* seen as having ethical stakes by most people, most of the time. In doing so, they also draw attention to how much of the "ordinary" is not taken to be ethical—and, indeed, how strange it can be to see ethics in so-called ordinary places. For debates around what "ordinary ethics" might mean, see Lempert 2013, 2015; Zigon 2014; Lambek 2015 Robbins 2016; Mattingly and Throop 2018; and Sidnell, Meudec, and Lambek 2019.

6. As we have already seen in chapter 2, such alignments of belonging and values were not, in practice, so automatic or easy to sustain. But there was, nonetheless, a widespread expectation that to ground one's ethics in one's community ties was the standard practice—the norm against which people like Francis defined their own ethics.

7. I call this meta-ethics only in the non-technical sense that it is a position on how ethics ought to be done, or an ethics of ethics, not to draw any connection to meta-ethics as a subfield of moral philosophy.

8. Lempert 2008; Jaffe 2009.

9. Du Bois 2007, 163.

10. In employing Du Bois's (2007) framework, I write about alignment/disalignment between subjects as if it were synonymous with alignment/disalignment between people. This is consistent with Du Bois's discussion, in which stances position distinct speakers in a conversation. However, one might also analyze subject positions as features of specific acts of stance-taking, such that one person/speaker could take different subject positions at different times. For example, I might criticize English medicine today but praise it tomorrow. An advantage of this approach is that alignment/disalignment between subject positions can describe not only relations between people but also relations among a person's various subject positions (that is, consistency or inconsistency). In other words, this allows one to differentiate between inter-speaker alignment and intra-speaker alignment. One might argue, in this vein, that radical activists were very much concerned with the subject-subject side of the triangle, but that they prioritized alignment among their own subject positions (intra-speaker) vis-à-vis alignment with the subject positions of others (inter-speaker).

11. For an in-depth analysis of the political economy of Indian tea plantations, including efforts to produce alternative "fair trade" tea, see Besky 2014.

12. To borrow the anthropologist Cheryl Mattingly's (2014) concept of "moral laboratories," radical environmental activists made the mundane activities of everyday life (especially eating and drinking, but also walking, going to a movie, and even breathing) into sites of intense ethical experimentation. But unlike the African American parents that Mattingly describes, this was not an "unwanted necessity" but very much a matter of "grand ambitions to change the world" (27) In this, some of them (including Hari) were self-consciously following the Gandhian tradition of life as moral experimentation (Gandhi 1954).

13. This can be seen as another instance of the politics of prefiguration, previously seen in the rain camp in chapter 1. Note, however, that the rain camp participants prefigured an ideal future primarily through their interactions with one another. In nature life, one's own body becomes the material for prefiguring change—caring for the self becomes a model of caring for the world.

14. While Sree Narayana Guru is most famous for his role in establishing the Sree Narayana Dharma Paripalana (SNDP), an organization for the reform and uplift of the Izhava caste, his teachings remain widely influential. Mohandas considered himself a disciple of Sree Narayana Guru, but rejected the SNDP's contemporary focus on the welfare of a particular group.

15. For more on the radical environmental activists' scalar ideologies of integration, see chapter 2.

16. Gandhi and Kumarappa 1954; Alter 2000; Lal 2000. In a move consistent with their broader politics of alternatives, Hari and others distinguished between their practice of nature cure and "naturopathy," which they saw as a professionalization and commercialization incompatible with the core principles of nature cure (cf. Alter 2015, 2016)

17. Alter 2020; Sheldon 2020.

18. Nature cure was central to the *swadeshi* (self-rule) practices at the heart of anti-colonial Gandhi's activism (Alter 2015; Sheldon 2020). But moral focus on diet also has deeper roots in India, including the importance given to dietary distinctions in caste hierarchy (Dumont 1970; Srinivas 1966; Marriott 1968; Busby 1997). Alter (2015) argues that the core principles of nature cure were also inherently compatible with Indian *Samkhya* philosophy, which is "based on an ontological understanding of ecological holism wherein the body and the environment interpenetrate one another on levels of sensory experience that cross-cut gross and subtle domains" (16). Similar notions of body-environment interpenetration are evident in the importance some activists, like Hari, gave to managing one's body in pursuit of environmental justice.

19. Barnett et al. 2005; Carrier 2008.

20. Elsewhere, I have analyzed the interactional dynamics by which such evaluative moves are rendered "sticky," compelling others to feel morally accountable for actions they may formerly have considered outside the purview of ethical concern (Mathias 2019). Once one is called to give an account of oneself (Butler 2005), justification only reinforces such implicit acts of moralization, making it hard to get back to seeing such mundane acts as drinking a soda or using soap as morally neutral.

21. The humor in this term echoed that of the term *buddhijīvi,* generally glossed as "intellectual," but also bringing to mind the fanciful notion of a species of *jīvi* ("creature") that has *buddhi* ("wisdom" or "intellect"). This term was also often spoken with a smirk and connoted a specific stereotype of Communist intellectual—a thin, heavily bearded man who carried a cloth bag over one shoulder, a figure both pretentious and comical.

22. In an analysis of a distinct but genealogically related dietary ethics among yoga practitioners in Mysore, anthropologist Jack Sidnell (2017, 172) describes a similarly propensity for dietary ethics to be socially divisive, noting that US and Canadian yoga practitioners, who aspire to vegetarianism, "express an acute sense of alienation from not being able to partake in the communal feast" during holidays featuring turkey or ham.

23. Adam Smith (2002 [1761], 21) suggested that such second-order evaluations were inherent to all evaluation, observing that even if one person only smiles at something that makes another person laugh, such a slight discordance

will entail judgment of each by the other (for empirical work that bears this out, see Pomerantz, 1984; Sidnell 2010). Regardless of whether this is true in all cases, the tendency to judge others' judgments was clearly heightened by the high moral stakes activists saw in otherwise mundane acts.

24. The term "camps" (in Malayalam *kūṭṭāymakaḷ,* but often used in English) refers to gatherings of environmentalists, usually overnight, for fellowship and mutual edification. Many of the camps Hari attended involved instruction in particular techniques of bodily care. See chapter 1 for an analysis of how camps were integral to the process of becoming an environmental activist.

25. According to Alter (2015, 2022), such personal stories of dramatic and even astonishing recovery are central to the epistemology of nature cure, which opposes the science of modern biomedicine not with "tradition" but with self-experimentation. Likewise, Hari's extension of lessons from his own bodily experiments to advising Sunathi is consistent with the broader transition of nature cure practitioners from healed to healer. Alter (2020) writes, "The best way to understand nature cure both globally and locally is with reference to publications by individuals who describe, in vivid and precise terms, how they cured themselves against all odds and then established clinics on the basis of personal experience" (113).

26. A powerful counter-example, however, can be seen in the trajectory of M.K. Gandhi's experiments with "alternatives," which so influenced Hari and other nature-cure practitioners. Gandhi's story has been told much better elsewhere than I can manage here, but a few convergences and contrasts are worth noting in passing. In one sense, we can see how Gandhi's dress, dietary habits, and sexual practices made him unusual or "out of place." But his strangeness was also what made him a powerful leader and moral exemplar, revered as a sort of organizer-saint (Fox 1989). Perhaps such fame was an as yet unrealized possibility for Hari as well.

But Hari's story is also arguably latent in Gandhi's. In an ethnographic and historical exploration of the notion of "cure" in India, Venkat (2021) revisits the story of the death of Gandhi's wife, Kasturba, who fell ill while the two of them were political prisoners of the British. Gandhi "arranged for a caravan of healers to visit Kasturba, including practitioners of both Ayurveda and nature cure" (212). But when Devdas, her youngest son, procured some imported penicillin, Gandhi "refused the treatment on his wife's behalf" (212). Venkat describes how various narrations of this event attempt to shore up the classic image of Gandhi as wise, benevolent, and insistent on justice. He includes in this Gandhi's criticism of allopathic medicine and modernity. But ultimately, he concludes that, "The accusation inherent in such a line of questioning is that Gandhi allowed his wife to die on principle. This is a vision of Gandhi as a dogmatic moralist" (213). This was, of course, the accusation against Hari as well. Much depends on how the story is told.

27. Modern homes (and other buildings) in Kerala are commonly built with a combination of stone (usually granite or red laterite) and cement mined from the Western Ghats mountain range that defines the state's eastern border. The anti-quarry protests described in chapter 1 and the conclusion were, therefore, also directed at Kerala's ongoing (and, arguably, decades-long) construction boom (Gopikuttan 1990), especially the construction of large apartment buildings.

28. Adarsh's appreciation for the family, at least in this moment, seems to share much with what Davé (2023) calls "indifference to difference" or "living with others in their otherness . . . the posture of side by side, rather than face to face" (1). Whereas Davé is primarily concerned with differences of species (but also, by extension, castes, genders, sexualities, abilities, and other human kinds), Adarsh's differences with his father are differences of ethical stance. And where such differences are concerned, indifference may become more difficult to distinguish from apathy, which would seem to run counter to both Adarsh's and Davé's projects.

## CHAPTER 4

1. Others have noted the remarkably high consumption of news in Kerala (Jeffrey 2000, 2009; Mithun 2012). I found that it was common for a middle-class family to subscribe to two or three daily newspapers, citing the need to counterbalance the biases of any one paper. Beginning in the early 2000s, twenty-four-hour Malayalam-language cable news channels began to proliferate, and there were about a dozen at the time of my research. Adarsh said that when the TV news had initially become popular, many expected it to supplant newspaper consumption. Instead, he suggested, it had only further stimulated the appetite for news across all media.

2. In other words, as the conflict with the reporter suggests, the Manamur protestors self-consciously saw their campaign as a series of what Francis Cody (2023, 4) calls "news events," in which news media (including both mass and social media) does not only represent events but "can be said to become an event in its own right." At the same time, as in the protest movements that Cody describes, such circulation was also understood to bring evidence before "the court of public opinion" (158). Thus, even as they worked within feedback loops of representation, Manamur protestors placed high stakes on the possibility that audiences would see the news not only as a representation of itself but also as a faithful representation of the injustices done to them.

3. Guha 1989. See also Govindrajan 2018.

4. This is not to be confused with object in the sense of "material thing" or "non-human." In my usage here, objects can be humans or immaterial things (such as ideas), so long as they are evaluated. In this sense, insofar as evaluating others is integral to ethics (see chapter 3), ethical subjects are also moral objects par excellence.

5. This chapter, thus, presents a different kind of answer to Gayatri Spivak's (2003) question "Can the subaltern speak?," a problem she took to be central to the Subaltern Studies group's project of accounting for the place of "the people" in history. Protestors in Manamur could speak for themselves, but being the people was not about making their own voices heard.

6. Anthropologists have challenged anthropocentrism (and the boundaries of their own discipline) by exploring subjectivity in nonhumans (for example, Locke 2017) or subjectivity as an effect of human interactions with nonhumans (Kirksey and Helmreich 2010; see also Lien and Pálsson 2021 for a review that situates this trend historically). Such work is important to understanding the

politics of environmental justice. For example, Martellozzo (2021) examines how forests flattened by windstorms in Italy act as "embedded witnesses" to the failures of a capitalist economy that drives both monocultural forestry and climate change (see also Vaughn and Fisher 2021). Yet there has been relatively little attention to humans as objects, which is often the other side of such stories. As this chapter shows, it is often in the moments when we need nonhumans (such as a smelly rice paddy) to bear witness that humans seek to position themselves as mere objects.

7. A District Collector is a government official in charge of revenue collection and administration of a district (a second-level Indian administrative division, a subdivision of states). As officers of the Indian Administrative Service, Collectors have a certain degree of autonomy from electoral politics. A Collector has the power to issue "stop work" orders for businesses operating within the district under her charge.

8. Deictics—such as English terms "I," "this," or "there"—are linguistic forms that specify the relevant context for an act of linguistic reference (Hanks 1993).

9. As Schutz (1967) argues, intersubjective coordination is not only a way of knowing some common object, but a way of knowing one another. Schutz describes this relationship using a metaphor of two people beside each other watching a bird in flight. They not only orient to the bird, they also orient to one another via the bird. Both people may not experience an object (such as the bird) in the world in the same way, but they nonetheless share a togetherness in their coordinated experiences of objects.

10. Keane (2016, 2) argues that ethical life depends on such "shared realities," which are not simply given but "built up, reshaped, or undermined, in time; between people."

11. Du Bois 2007.

12. Such tours of pollution, or "toxic tours," are a common tactic in environmental justice movements outside of India as well (Pezzullo 2009; Avenell 2012).

13. Mazzarella 2006; Meyer 2011; Keane 2008.

14. In dismissing the local activists' efforts to mediate directness, the researcher appears to be practicing a version of scientific objectivity, grounded in an ethos of detachment and impartiality (Daston 1999). As Daston and Galison (2007) document, there are many ideologies and practices of objectivity, but they share an aim of purifying communication of interference from the biases, perspectives, desires, or other particular qualities (in short, the *subjectivity*) of subjects.

Thus, the persuasive tactics I describe in this chapter are all practices of objectivity, but each exploits a different ideology about how objects can be made to communicate clearly. Notably, Action Council members wielded multiple modes of objectivity simultaneously; they were not stuck "inside" a particular ontology. But they nonetheless fully lived and inhabited each mode: cancers were feared and rage was deeply felt.

By contrast, consider anthropologist Kim Fortun's (2001, 37) description of the efforts of activists in northern India to deploy heterogeneous logics as they advocated for survivors of the 1984 Bhopal gas spill: "The challenge is not to stand outside established systems, but to find places to work within them—finding repose in none, recognizing that neither poetry nor law is sufficient in itself. . . . [T]he challenge is to move between different places of articulation, recognizing that no one form can tell the whole story." Local activists in Manamur also had to move between rhetorical forms, but they could not do so easily because rhetorical forms was so closely bound to forms of life.

15. At any given time, there were several legal cases in the courts, and Vijayan often spoke optimistically about the potential for one of these cases to deliver victory. Yet, like the government research that called for further research, the legal process always seemed to drag out longer than expected, with potential victories always just beyond the next rise. This process was also largely beyond the scope of my fieldwork. Vijayan and other Action Council members understandably felt it would be problematic for me to attend court sessions or meetings with legal counsel. Like many of the local protestors, I witnessed the campaign's legal strategy only via the updates that came back through Vijayan and other Action Council leaders—and the best of these updates only held out the hope that a game-changing legal victory might be somewhere up ahead.

16. See, for example, Reno (2011); Tesh (2000); Checker (2005).

17. As Ottinger and Cohen (2011, 6) write, environmental justice movements encounter scientists and engineers both as experts "who are deeply implicated in creating and maintain environmental injustices" and as potential allies whose expertise may be leveraged to challenge or redress injustices (Pellow and Brulle 2005; Fischer 2000).

18. Tesh 2000.

19. Similarly, Choy (2005) writes that environmental activists' uses of scientific expertise should not be seen simply as deployments of "universalizing" ways of knowing. Rather, he argues, "to be credible, expertise must bear universalizing and particularizing marks simultaneously" (6).

20. Isaac, Franke, and Parameswaran 1997; Jaffry et al. 1983.

21. Most notably, the conservation movement to prevent a dam that would have flooded Kerala's Silent Valley (Manjusha 2016).

22. Like many Catholic priests of his generation in Kerala, the founder of *Jananīti* was inspired by liberation theology. This led to involvement with environmental justice protests and, ultimately, to his expulsion from the priesthood.

23. The influence of the *Jananīti* report had profound impacts on the experience of pollution. In chapter 2, I note that one Action Council member, Vinod, confided in me that he had been ill and felt certain that he had cancer. This experience was shaped by the *Jananīti* report and other research that had found evidence of carcinogens in the factory's pollution. In this chapter, I largely analyze the suffering of Manamur protestors in terms of rhetorical strategy and performance. But this is in no way to diminish experiential and affective realities of living with "the smell."

24. Solidarity was the official youth wing of Jamaat-e-Islami in Kerala. Jamaat-e-Islami is a Muslim political party with relatively low membership in Kerala but a strong media presence via its daily newspaper, weekly magazine, and news channel. The work of Solidarity often paralleled the people's protest

activism described in chapter 1, and a few Solidarity activists were part of the *Kēraḷīyam* social circle. However, many activists in the latter group were highly skeptical of Solidarity, viewing its people's protest work as a thinly veiled effort to gain a foothold for Jamaat-e-Islami.

25. Latour 2000, 115. Writing about the practice of social science, Latour (2005, 79) says, "Objects, by the very nature of their contact with humans, quickly shift from being mediators to intermediaries, counting for one or nothing, no matter how internally complicated they might be. That is why specific tricks must be invented to *make them talk* . . ." (italics his). In this chapter, I extend this point to practices of ethical objectivity, showing how activists seek to make pollution speak for itself. At the same time, however, I show that they do this by putting themselves in an object position, making their protest a mere "intermediary" rather a "mediator" in Latour's terms—that is, a faithful echo of the pollution's voice. This is one thing ideologies of immediacy are good for.

26. In the late nineteenth century, the practice of caste unapproachability—in which those from oppressed castes were forbidden to come within a certain distance, or sometimes even within sight, of members of higher castes—was widespread in Kerala (Namboodiri 1999; Harikrishnan 2023). Some were prohibited from allowing themselves to be seen by the highest castes and could only travel roads with great difficulty, calling ahead of themselves continually to warn any who might be coming the other way. Oppressed castes would then step down from the road to let higher castes pass. Thus, even when differently casted bodies did travel the roads, the narrow linearity of the thoroughfare did not bring these bodies together. When the oppressed castes began to organize in opposition to caste discrimination in the early twentieth century, it is not surprising that marches were central to their politics; claiming the road was itself one of the most powerful assertions of rebellion (Lemercinier 1984; Nisar and Kandasamy 2007). For a historical account of how roads in Kerala became public spaces, see Harikrishnan (2023).

27. In her ethnography of crowd politics in Bangladesh, Nusrat S. Chowdhury (2019) writes of streets as preferred public spaces for the practice of politics in South Asia. She writes that "the street is a public thing in that it gathers people together materially and symbolically, even when they are divisive" (10). The modes of publicness offered by South Asian streets, she argues, offer new possibilities for rethinking predominant notions of the public sphere as a space of disembodied, detached deliberation (see also Cody 2015).

28. In this sense, the Manamur Action Council had a tactical advantage that earlier environmental justice movements, including the famous Chipko movement, had not had. Guha (1989) suggests that the success of that movement's tree-hugging tactics in international media was largely fortuitous—more an effect of its meanings for Western audiences than for activists themselves. But due to technological advances in mass media, not to mention the advent of social media, the Manamur Action Council was able to quickly ascertain how distant audiences were representing their activities and adjust accordingly. Thus, while media coverage of their protest was still largely beyond their control, they could nonetheless respond to it with great agility.

29. Laclau 2005.

30. Trisal (n.d.) describes the challenges of sustaining such widespread collaboration in *hartals* in Kashmir against the Indian state. Organizers of the *hartals* published calendars specifying "the time at which the strike was to begin and end each day as well as its duration and intensity throughout any given week. . . . The revolution it turns out, would be scheduled" (1–2). Contrast this with the semiotics of spontaneity employed in many protest movements, including by protestors in Manamur.

31. Compare also Chowdhury's (2019) ethnography of crowds in Bangladesh, in which the aim is also "to take over the street," to explore "when and how a numerical minority of individuals physically gathered in a public space can be understood to speak and act on behalf of a superior but forever disembodied entity called "the people" (9, 29). This chapter explores this same question. But because the street politics of people's protests cannot hope, usually, to take over streets, it offers somewhat different answers. The performative tactics of people's protests should, however, be seen as complementary—and even symbolically dependent upon—the crowd politics of parties and large social organizations.

32. M. Chatterjee 2016. Genevieve Lakier (2014) writes that *hartals* (known in some areas as bandhs) called by Maoist revolutionaries in Nepal "operate not merely to *represent* but to *constitute* power by demonstrating their ability to coerce participation and in so doing establish a kind of quasi-sovereign, albeit temporary, control over the public sphere." Of course, as Lakier also documents and I discuss here, *hartals/bandhs* are also thoroughly representational—they are public spectacles. But this spectacle depends on a coordination of bodies, both willing and coerced, that is often violent. This spectacle of coordination may contrast sharply with the semiotics of spontaneity I describe in the street politics of Kerala's people's protests. Of roadblocks led by politicians in Mumbai, Bjorkman (2015, 158) writes, "The more orchestrated and staged an event appears . . . the more persuasive and real it is as a sign of the orchestrator's authority."

33. As such, *hartals/bandhs* called by political parties in Kerala should not be seen as challenging state sovereignty, per Chakrabarty's (2007) argument. Rather, as Björkman (2015) argues, such forms of street protest can bolster the legitimacy of the state by making demands for state intervention or services. Likewise, Mitchell (2023) groups *hartal* with a wide repertoire of street politics that take the state as a primary addressee and, thus, interpellate the state as a legitimate arbiter of competing claims. *Hartal* can be one way of getting unresponsive state actors to listen; especially for marginalized populations, so-called uncivil tactics can be a way of amplifying their voices (Mitchell 2023). As a tactic of opposition parties, *hartals* are arguably aimed at delegitimizing the current government (vs. state) and claiming, by way of contrast, the legitimacy of opposition protest as an expression of public sentiment.

34. For an autobiographical account of the coining of the term, see Gandhi (1928). Guha (2013) discusses the uptake of *satyagraha* in Indian environmental movements. For broader discussion and commentary, see Alter 2000, Skaria 2016. As Trisal (n.d.) argues, for Gandhi *hartal* was also one tactic or manifestation of the broader practice of *satyagraha*, likewise employing a display of self-sacrifice to persuade.

35. Alter 2000; Laidlaw 2005; Skaria 2010.

36. Busby 1997; Holdrege 1998; Marriott and Inden 2011.

37. Vijayalekshmy (1999) recounts a legendary case in which the locals of Matilakam protested the construction of walls around the town. When non-Brahmin residents lay across the site to obstruct construction, the rulers built the wall over their bodies. But when Brahmins conducted a group fast, "society could not bear this" (415). The curse of the fast sowed discord among the rulers, and in the ensuing conflict the rulers' houses were destroyed, the walls razed, and the town burned.

38. Performing sincerity could, in turn, lead to greater fame and public esteem. Indeed, some activists in Manamur suggested that it was in pursuit of such rewards that Stevenson had been so eager to inaugurate the hunger strike.

39. Bargu 2014, 16; compare Laidlaw (2005).

40. Rancière 1999, 9.

41. Foucault 1978, 138.

42. Halliburton (2002) argues for the salience of multiple levels of "rarification" between material and immaterial aspects of persons in Kerala. While "consciousness" (*bōdham*) is a relatively disembodied motive for people's protest activism (see chapter 2), *vikāram* engages the body, yet its force as a motive comes from its capacity to overflow local mind-body distinctions.

43. Polletta 2006.

44. At the time of my research, this was the state of the campaign for reparations from pollution by a Coca-Cola plant in Plachimada, Kerala, which had been shut down in 2005 (Sreemahadevan 2008). In her ethnography of activism following the Bhopal gas disaster, Fortun (2001) writes, "The seeming simplicity of the call for justice is, however, deceptive. Gas victims and their allies seem to know this—often emphasizing their awareness that justice itself will always be deferred, that there is no way to fully rehabilitate Bhopal" (352).

45. The record of some other people's protests in Kerala suggests, however, that such local uprisings against industrial pollution are far from assured. Examining the case of Eloor, one of Kerala's most polluted areas, Devika and Narayanan (2019) explore why most residents are resistant to protest despite longstanding experiences of respiratory illness and other health effects. They argue that Eloor reseidents remain nostalgic for the vision of burgeoning industry that originally drew them there—what Devika and Narayanan call an "industrial heterotopia"— with its promises of social mobility and development. As shown in this chapter and in chapter 2, much work is required to produce localness as a category of environmental resistance.

## CONCLUSION

1. This embarrassing incident was the result of a misunderstanding of the meaning of a plate of food left on the kitchen counter at night. For discussion of the capacity of gifts to "become lost to those whom they would serve," see Keane (1997, 91–93).

2. Singer's "expanding circle" argument (see introduction), falls within the utilitarian tradition, which offers some of the most full-throated examples of

this emphasis (classic statements include Bentham 1789; Godwin 1793). But critics of this emphasis have often taken Immanuel Kant's moral philosophy as their chief target. For example, Bernard Williams (1981, 2) writes that, for Kant, "the moral point of view is specially characterized by its impartiality and its indifference to any particular relations to particular persons" (see also MacIntyre 1966; Tronto 1993). Even Barbara Hermann (1993, 184), whose account of Kantian ethics pushes back on these criticisms and makes room for social attachment and care, acknowledges there are good reasons to see Kant's philosophy as "the standard model of an impartial ethical system." Yet shades of Triangle B in can be found, in more subtle forms, well beyond the utilitarian and Kantian traditions. For example, Tronto (1993) sees linkages between Adam Smith's moral philosophy and the feminist ethic of care discussed in chapter 2. But even Adam Smith, whose moral philosophy begins from consideration of the fundamental interdependence of values and relationships, still arguably leans toward Triangle B by deriving, through social interaction and fellow-feeling, the notion of an "impartial spectator" as moral arbiter (Smith 2002, 150). Likewise, Tronto (1993, 142) argues that partiality and parochialism continue to be a challenge for ethics that emphasize "engagement with the concrete, the local, the particular."

3. For discussions of the influence of virtue ethics in the anthropology of ethics, see Mattingly 2012; Mattingly and Throop 2018.

4. See introduction for discussion. For a critical review, see Venkatesan 2023.

5. Such transformations are not unusual in environmentalist circles. Of her fieldwork among Friends of the Earth activists, Gatt (2017) writes, "I increasingly began monitoring my choices and actions according to a vague notion of sustainability. Having been shown a film about the violent workings of the Coca-Cola company in Colombia, I began to choose not to drink Coca-Cola (the choice was an internal battle every time I bought a drink). Later still, after more reflection and discussion, the practice turned into not buying or drinking fizzy drinks made by multinational corporations." The American naturalist Aldo Leopold (1966 [1953], 183) said simply, "One of the penalties of an ecological education is that one lives alone in a world of wounds." My own less-than-willing transformation, of course, speaks to the fact that ethical life is never, really, lived alone.

6. "Quarry mafia" did not necessarily refer exclusively to organized crime, but was used more broadly to reference the shadowy business interests that were understood to control the quarrying economy. Quarry mafias were understood both to often operate illegally and to use bribes and threats to exert illicit influence over quarrying governance and policy. For a study of the quarry "mafia" that situates its business practices within the broader politics of development in Kerala, see essays in Devika et al., 2022.

7. The anthropologist Saba Mahmood (2005), in her study of piety movements among Muslim women in Cairo, describes a process of ethical formation that traverses boundaries between internal and external much more profoundly and completely—not only introducing accountability into a conversation, but fundamentally transforming opinions and desires. She describes how, in the self-understanding of these women, "submission to certain forms of (external)

authority is a condition for the self to achieve its potentiality" (149). Mahmood's argument about the entanglement of internal and external forces is very much in line with my own arguments in the previous chapters. But Ahmed's engagement with the norms of the activists we studied was very different from the process of ethical formation Mahmood describes. For example, while Ahmed often expressed admiration for Faiza and Adarsh's activist commitment and seemed to desire their approval, I saw no indication that he actively cultivated this desire for approval (nor fear of disapproval), nor was it clear whether or how his fear of being "found out" by Faiza or Rajendran shaped his own commitments. If their voices were internal to his moral dialogue, this was not necessarily because he had welcomed them in.

8. Many of the largest and most elaborate houses I saw in Kerala were owned by people working overseas, especially in the Gulf. These were often described, by activists and non-activists alike, as sitting empty.

9. This process corresponds closely with Adam Smith's (2002 [1761]) account of ethical formation. While Durkheim (1961) saw ethical formation as the work of society upon the individual, and Foucault (1990) focused on the work of the self upon the self, Smith's moral philosophy draws attention to the mutual influence of evaluating subjects upon one another and, especially, on the self-forming process of seeing oneself through the evaluative eyes of one's interlocutors (see chapter 3). Ahmed's imagined dialogue is neither a form of social discipline nor a practice of freedom, but a nod to each as he works his way through a world riddled with evaluation.

10. A *pavan* is about eight grams. Such gold is commonly used as collateral for obtaining loans should the married couple encounter financial duress.

## APPENDIX

1. I began studying Malayalam under the tutelage of a local government clerk, using the classic textbook composed by Rodney Moag (1980). I continued with one year of study at the University of Chicago, followed by two summers and one year of study with the American Institute of Indian Studies program in Thiruvananthapuram.

2. Emerson, Fretz, and Shaw 1995.

3. Ajayan 2017; McGarr 2014.

4. Zuckerman 2023.

5. In a powerful essay, Shah (2022) describes how her own purposes in writing emerged, in part, from her experiences doing fieldwork among Naxalite activists. Likewise, though I do not see this book as activist anthropology in the usual sense (see introduction), ongoing dialogue with activists about their purposes led me toward a sense of my own purpose in writing about them—above all, to a desire to do justice to the stakes and challenges that they saw in their own lives.

# Bibliography

Abu-Lughod, Lila. 1991. "Writing Against Culture." In *Recapturing Anthropology: Working in the Present,* edited by Richard G. Fox, 137–62. Santa Fe, NM: School of American Research Press.

Ahmann, Chloe. 2018. "It's exhausting to create an event out of nothing": Slow Violence and the Manipulation of Time." *Cultural Anthropology* 33, no. 1: 142–71.

———. 2020. "Atmospheric Coalitions: Shifting the Middle in Late Industrial Baltimore." *Engaging Science, Technology, and Society* 6: 462–85.

Aiyappan, A. 1965. *Social Revolution in a Kerala Village: A Study in Culture Change.* London: Asia PublishingHouse.

Aiyer, Ananthakrishnan. 2007. "The Allure of the Transnational: Notes on Some Aspects of the Political Economy of Water in India." *Cultural Anthropology* 22, no. 4: 640–58.

Ajayan, T. 2017. "Dismissal of the First Communist Ministry in Kerala and the Role of Extraneous Agencies." *The South Asianist Journal* 5, no. 1.

Alinsky, Saul David. 1971. *Rules for Radicals: A Practical Primer for Realistic Radicals.* New York: Random House.

Alter, Joseph S. 2015. "Nature Cure and Ayurveda: Nationalism, Viscerality and Bio-ecology in India." *Body & Society* 21, no. 1: 3–28. doi: 10.1177/1357034X14520757.

———. 2020. "From Lebensreform to Swadeshi: Vithal Das Modi and the Development of Nature Cure in India." *Asian Medicine* 15, no. 1: 107–32.

———. 2000. *Gandhi's Body: Sex, Diet, and the Politics of Nationalism.* Philadelphia: University of Pennsylvania Press.

Alter, Joseph Stewart, and Chandrashekar Sharma. 2016. "Nature Cure Treatment in the Context of India's Epidemiological Transition." *Journal of*

*Integrative Medicine* 14, no. 4: 245–54. doi: https://doi.org/10.1016/S2095-4964(16)60265-0.

Álvarez, Lina, and Brendan Coolsaet. 2020. "Decolonizing Environmental Justice Studies: A Latin American Perspective." *Capitalism Nature Socialism* 31, no. 2: 50–69.

Ambrosone, Ellen. 2022. "Language Reform in 19th-Century Kerala." *Oxford Research Encyclopedia of Asian History.* https://oxfordre.com/asianhistory/view/10.1093/acrefore/9780190277727.001.0001/acrefore-9780190277727-e-416.

Ambrosone, Ellen A., Laura A. Ring, and Mara L. Thacker. 2023. "Collections, Care, and the Collective: Experiments in Collaborative Fieldwork in Area Studies Librarianship." *IFLA Journal* 49, no. 1: 39–51.

Appadurai, Arjun. 1990. "Disjuncture and Difference in the Global Cultural Economy." *Theory, Culture & Society* 7, no. 2–3: 295–310.

Arunima, G. 2003. *There Comes Papa: Colonialism and the Transformation of Matriliny in Kerala,* Malabar, c. 1850–1940. New Delhi: Orient Longman.

———. 2006. "Imagining Communities–Differently: Print, Language and the (Public Sphere) in Colonial Kerala." *The Indian Economic & Social History Review* 43, no. 1: 63–76. doi: 10.1177/001946460504300103.

Asher, R. E, and T. C. Kumari. 1997. *Malayalam.* New York: Routledge.

Attfield, Robin. 2013. "Climate Change, Environmental Ethics, and Biocentrism." In *Climate Change and Environmental Ethics*, edited by Ved P. Nanda, 31–42. New Brunswick, NJ: Transaction Publishers.

Avenell, Simon. 2012. "From Fearsome Pollution to Fukushima: Environmental Activism and the Nuclear Blind Spot in Contemporary Japan." *Environmental History* 17, no. 2: 244–76.

Bargu, Banu. 2014. *Starve and Immolate: The Politics of Human Weapons.* New York: Columbia University Press.

Barnett, Clive, Paul Cloke, Nick Clarke, and Alice Malpass. 2005. "Consuming Ethics: Articulating the Subjects and Spaces of Ethical Consumption." *Antipode* 37, no. 1: 23–45.

Batavia, Chelsea, Jeremy T. Bruskotter, Julia A. Jones, and Michael Paul Nelson. 2020. "Exploring the Ins and Outs of Biodiversity in the Moral Community." *Biological Conservation* 245: 108580. doi: https://doi.org/10.1016/j.biocon.2020.108580.

Baviskar, Amita. 1997. "Tribal Politics and Discourses of Environmentalism." *Contributions to Indian Sociology* 31, no. 2: 195–223.

———. 2004. *In the Belly of the River: Tribal Conflicts over Development in the Narmada Valley.* 2nd ed. New York: Oxford University Press.

———. 2005. "Red in Tooth and Claw? Looking for Class in Struggles Over Nature." In *Social Movements in India: Poverty, Power, and Politics*, edited by Raka Ray and Mary Fainsod Katzenstein, 161–78. New York: Rowman and Littlefield.

———. 2020a. "Cows, Cars and Cycle-Rickshaws: Bourgeois Environmentalists and the Battle for Delhi's Streets." In *Elite and Everyman*, edited by Amita Baviskar and Raka Ray, 391–418. New Delhi: Routledge.

———. 2020b. *Uncivil City: Ecology, Equity and the Commons in Delhi.* New Delhi: Sage.

Bayly, C. A. 2009. "The Indian Ecumene: An Indigenous Public Sphere." In *The Indian Public Sphere: Readings in Media History,* edited by Arvind Rajagopal, 49–64. New Delhi: Oxford.

Bellacasa, María Puig de la. 2017. *Matters of Care: Speculative Ethics in More Than Human* Worlds. Minneapolis: University of Minnesota Press.

Benedict, Ruth. 1934. *Patterns of Culture.* Boston: Houghton Mifflin.

Benford, R. D., and D. A. Snow. 2000. "Framing Processes and Social Movements: An Overview and Assessment." *Annual Review of Sociology* 26: 611–39.

Bentham, Jeremy. 1907 [1789]. *An Introduction to the Principles of Morals and Legislation.* Oxford: Clarendon Press.

Berglund, Henrik, and Sofia Helander. 2015. "The Popular Struggle against Coca-Cola in Plachimada, Kerala." *Journal of Developing Societies* 31, no. 2: 281–303.

Besky, Sarah. 2014. *The Darjeeling Distinction: Labor and Justice on Fair-Trade Tea Plantations in India.* Berkeley, CA: University of California Press.

Binoy, P. 2014. "Struggling against Gendered Precarity in Kathikudam, Kerala." *Economic And Political Weekly* 49, no. 17: 45–52.

Björkman, Lisa. 2015. "The Ostentatious Crowd: Public Protest as Mass-Political Street Theatre in Mumbai." *Critique of Anthropology* 35, no. 2: 142–65.

Boggs, C. 1977. "Marxism, Prefigurative Communism and the Problem of Workers' Control." *Radical America* 6 (Winter): 99–122.

Boyd, David R. 2017. *The Rights of Nature: A Legal Revolution That Could Save the World.* Toronto, ON: ECW Press.

Breines, W. 1989. *Community and Organization in the New Left, 1962–1968: The Great Refusal.* New Brunswick: Rutgers University Press.

Brennan, Andrew. 1984. "The Moral Standing of Natural Objects." *Environmental Ethics* 6, no. 1: 35–56.

Brennan, Andrew, and Yeuk-Sze Lo. 2002. "Environmental Ethics." *Stanford Encyclopedia of Philosophy.* https://plato.stanford.edu/entries/ethics-environmental.

Bryant, Bunyan. 2003. "History and Issues of the Environmental Justice Movement." In *Our Backyard, A Quest for Environmental Justice*, edited by G. R. Visgilio and D. M. Whitelaw, 3–23. Lanham, MD: Rowman and Littlefield.

Buch, Elana D. 2015. "Anthropology of Aging and Care." *Annual Review of Anthropology* 44: 277–93.

Bullard, Robert D. 1990. *Dumping in Dixie: Race, Class, and Environmental Quality.* Boulder, CO: Westview.

Burningham, Kate 2000. "Using the Language of Nimby: A Topic for Research, Not an Activity for Researchers." *Local Environment* 5, no. 1: 55–67.

Busby, Cecilia. 1997. "Permeable and Partible Persons: A Comparative Analysis of Gender and Body in South India and Melanesia." *Journal of the Royal Anthropological Institute* 3, no. 2: 261–78.

Butler, Judith. 2005. *Giving an Account of Oneself.* New York: Fordham University Press.

Callicott, J. Baird. 1987. "Conceptual Resources for Environmental Ethics in Asian Traditions of Thought: A Propaedeutic." *Philosophy East and West* 37 (2): 115–30.

———. 2013. "Toward an Earth Ethic: Aldo Leopold's Anticipation of the Gaia Hypothesis." In *Climate Change and Environmental Ethics*, edited by Ved P. Nanda, 17–30. New Brunswick, NJ: Transaction Publishers.

Canovan, Margaret. 2005. The People. Cambridge, UK: Polity.

Carr, E. Summerson, and Michael Lempert. 2016. "Introduction: Pragmatics of Scale." In *Scale: Discourse and Dimensions of Social Life,* edited by E. S. Carr and M. Lempert, 1–24. Oakland: University of California Press.

Carr, E. Summerson, and Brooke Fisher. 2016. "Interscaling Awe, De-escalating Disaster." In *Scale: Discourse and Dimensions of Social Life*, edited by E. Summerson Carr and Michael Lempert, 133–58. Oakland: University of California Press.

Carrier, James G. 2008. "Think Locally, Act Globally: The Political Economy of Ethical Consumption." In *Hidden Hands in the Market: Ethnographies of Fair Trade, Ethical Consumption, and Corporate Social Responsibility,* edited by Geert De Neve, Luetchford Peter, Jeffrey Pratt and Donald C. Wood, 31–51. Bingley, UK: Emerald Group Publishing Limited.

Census of India. 2011. Primary Census Abstract—By Religion. Accessed 08/05/2023. https://censusindia.gov.in/census.website/data/census-tables.

Center for Development Studies. 1975. *Poverty, Unemployment and Development Policy : A Case Study of Selected Issues with Reference to Kerala.* New York: United Nations.

Chakrabarty, Dipesh. 2007. "'In the Name of Politics': Democracy and the Power of the Multitude in India." *Public Culture* 19, no. 1: 35–57.

Chandran, Civic. 2012. "Namukkentāṉ liṯṯal māsikakaḷillattat." *Mathrubhumi* 90, no. 42: 14–15.

Chatterjee, Moyukh. 2016. "Bandh Politics: Crowds, Spectacular Violence, and Sovereignty in India." *Distinktion: Journal of Social Theory* 17, no 3: 294–307.

Chatterjee, Partha. 1993. *The Nation and Its Fragments: Colonial and Postcolonial Histories.* Princeton, NJ: Princeton University Press.

———. 2004. *The Politics of the Governed: Reflections on Popular Politics in Most of the World.* New York: Columbia University Press.

———. 2020. *I Am the People: Reflections on Popular Sovereignty Today.* New York: Columbia University Press.

Chaturvedi, Ruchi. 2011. "'Somehow it Happened': Violence, Culpability, and the Hindu Nationalist Community." *Cultural Anthropology* 26, no. 3: 340–62.

———. 2015. "Political Violence, Community and its Limits in Kannur, Kerala." *Contributions to Indian Sociology* 49, no. 2: 162–87.

Checker, Melissa. 2005. *Polluted Promises: Environmental Racism and the Search for Justice in a Southern Town.* New York: New York University Press.

———. 2011. "Wiped Out by the 'Greenwave': Environmental Gentrification and the Paradoxical Politics of Urban Sustainability." *City & Society* 23, no. 2: 210–29.

Chowdhury, Nusrat Sabina. 2019. *Paradoxes of the Popular: Crowd Politics in Bangladesh:* Stanford University Press.

Choy, Timothy K. 2005. "Articulated Knowledges: Environmental Forms after Universality's Demise." *American Anthropologist* 107, no. 1: 5–18.

Chua, Jocelyn Lim. 2014. *In Pursuit of the Good Life: Aspiration and Suicide in Globalizing South India.* Berkeley: University of California Press.

Cody, Francis. 2013. *The Light of Knowledge: Literacy Activism and the Politics of Writing in South India.* Ithaca, NY: Cornell University Press.

———. 2015. "Populist Publics: Print Capitalism and Crowd Violence beyond Liberal Frameworks." *Comparative Studies of South Asia, Africa and the Middle East* 35, no. 1: 50–65.

———. 2023. *The News Event: Popular Sovereignty in the Age of Deep Mediatization.* Chicago: University of Chicago Press.

Cohn, Bernard S. 1996. *Colonialism and Its Forms of Knowledge: The British in India.* Princeton, NJ: Princeton University Press.

Cole, Luke W., and Sheila R. Foster. 2001. *From the Ground Up: Environmental Racism and the Rise of the Environmental Justice Movement.* New York: New York University Press.

Collins, Patricia Hill. 2000. *Black Feminist Thought: Knowledge, Consciousness, and the Politics of Empowerment.* 2nd ed. New York: Routledge.

Daniel, E. Valentine. 1984. *Fluid Signs: Being a Person the Tamil Way.* Berkeley: University of California Press.

Das, Veena. 1998. "Wittgenstein and Anthropology." *Annual Review of Anthropology* 27, no. 1 171–95.

———. 2010. "Engaging the Life of the Other: Love and Everyday Life." In *Ordinary Ethics: Anthropology, Language, and Action,* edited by Michael Lambek, 1–38. New York: Fordham University Press.

———. 2012. "Ordinary Ethics." In *A Companion to Moral Anthropology,* edited by D. Fassin, 133–49. Oxford: Wiley Blackwell.

Daston, Lorraine. 1999. "Objectivity and the Escape from Perspective." In *The Science Studies Reader,* edited by Mario Biagioli, 110–23. New York: Routledge.

Daston, Lorraine, and Peter Galison. 2007. *Objectivity.* New York: Zone Books.

Dave, Naisargi N. 2012. *Queer Activism in India: A Story in the Anthropology of Ethics.* Durham, NC: Duke University Press.

———. 2015. "Love and Other Injustices: On Indifference to Difference." Humanities Futures, Franklin Humanities Institute Interdepartmental Seminar, March 15, 2015.

Davé, Naisargi N. 2023. *Indifference: On the Praxis of Interspecies Being.* Durham, NC: Duke University Press.

Deleuze, Gilles. 2019. "Cinema II: the Time-image." In *Philosophers on Film from Bergson to Badiou: A Critical Reader,* edited by Christopher Kul-Want, 177–99. New York: Columbia University Press.

Descombes, Vincent. 2016. *Puzzling Identities.* Translated by Steven Adam Schwartz. Cambridge: Harvard University Press.

Devika, J. 2008. "Rethinking 'Region': Reflections on History-Writing in Kerala." *Contemporary Perspectives* 2, no. 2: 246–64.

———. 2007. "'A people united in development': Developmentalism in Modern Malayalee Identity." Working Paper 386. Thiruvananthapuram: Centre for Development Studies.

———. 2010. "Egalitarian Developmentalism, Communist Mobilization, and the Question of Caste in Kerala State, India." *Journal of Asian Studies* 69, no. 3: 799–820.

Devika, J., S. Mohanakumar, and Archana Ravi. 2022. "Cronyism, Development, and Citizenship: A Study of the Effects of Quarrying in Pallichal Panchayat, Thiruvananthapuram." In *CDS Monograph Series: Ecological Challenges and Local Self-Government Responses*. Thiruvananthapuram, India: Centre for Development Studies.

Devika, J, and N.C. Narayanan. 2019. "The Local as Industrial Heterotopia: Making Sense of the Denial of Environmental Destruction at Eloor." In *Why Do People Deny Environmental Destruction? The Pollution of the Periyar at Eloor and Local-level Responses*. Thiruvananthapuram, India: Centre for Development Studies.

Devine-Wright, P. 2011. "Public Engagement with Large-Scale Renewable Energy Technologies: Breaking the Cycle of Nimbyism." *WIREs Climate Change* 2, no. 1: 19–26.

Dhanesh, Ganga S. 2010. "Kerala: God's Own Country." In *Public Relations Cases: International Perspectives,* edited by Danny Moss, Melanie Powell and Barbara DeSanto, 56-74. New York: Routledge.

Diehm, Christian. 2010. "Minding Nature: Val Plumwood's Critique of Moral Extensionism." *Environmental Ethics* 32, no. 1: 3-16.

Dirks, Nicholas B. 2001. *Castes of Mind: Colonialism and the Making of Modern India*. Princeton, NJ: Princeton University Press.

Dominelli, Lena. 2012. *Green Social Work: From Environmental Crises to Environmental Justice*. Cambridge, UK: Polity.

Douglas, Mary. 1966. *Purity and Danger: An Analysis of Concepts of Pollution and Taboo*. London: Routledge.

Dowie, Mark. 1996. *Losing Ground: American Environmentalism at the Close of the Twentieth Century*. Cambridge, MA: MIT Press.

Drengson, Alan. 2008. "The Life and Work of Arne Naess: An Appreciative Overview by Alan Drengson." In *The Ecology of Wisdom: Writings by Arne Naess,* edited by A. Naess, A. Drengson, & B. Devall, 3–44. Berkeley, CA: Counterpoint.

Du Bois, John W. 2007. "The Stance Triangle." In *Stancetaking in Discourse: Subjectivity, Evaluation, Interaction,* edited by Robert Englebretson, 139–82. Philadelphia: John Benjamins Publishing.

Dumont, Louis. 1970. *Homo Hierarchicus: An Essay on the Caste System*. Chicago: University of Chicago Press.

Durkheim, Émile. 1915. *The Elementary Forms of the Religious Life*. Translated by Joseph Ward Swain. New York: Macmillan.

———. 1951. *Suicide, a Study in Sociology*. New York: Free Press.

———. 1961 [1925]. *Moral Education: A Study in the Theory and Application of the Sociology of Education*. New York: Free Press of Glencoe.

———. 1984 [1893]. *The Division of Labor in Society*. Translated by W.D. Halls. New York: Free Press.
Edelman, Marc. 2001. "Social Movements: Changing Paradigms and Forms of Politics." *Annual Review of Anthropology* 30: 285–317.
Emerson, Robert M., Rachel I. Fretz, and Linda L. Shaw. 1995. *Writing Ethnographic Fieldnotes*. Chicago: University of Chicago Press.
Engel, Mylan Jr. 2008. "Ethical Extensionism." In *Encyclopedia of Environmental Ethics and Philosophy*, edited by J. Baird Callicott and Robert Frodeman, 396–98. Detroit, MI: MacMillan Reference.
Eranti, Veiko. 2017. "Re-Visiting Nimby: From Conflicting Interests to Conflicting Valuations." *The Sociological Review* 65, no. 2: 285–301.
Escobar, Arturo. 1995. *Encountering Development: The Making and Unmaking of the Third World*. Princeton: Princeton University Press.
Faubion, James D. 2011. *An Anthropology of Ethics*. Cambridge, UK: Cambridge University Press.
Ferguson, James. 1994. *The Anti-Politics Machine: "Development," Depoliticization, and Bureaucratic Power in Lesotho*. Minneapolis: University of Minnesota Press.
Fischer, Frank. 2000. *Citizens, Experts, and the Environment*. Durham, NC: Duke University Press.
Floyd, Juliet. 2016. "Chains of Life: Turing, Lebensform, and the Emergence of Wittgenstein's Later Style." *Nordic Wittgenstein Review* 5, no. 2: 7–89.
———. 2018. "*Lebensformen:* Living Logic." In *Language, Form(s) of Life, and Logic*, edited by Martin Christian, 59–92. Berlin: De Gruyter.
———. 2020. "Wittgenstein on Ethics: Working through Lebensformen." *Philosophy & Social Criticism* 46, no. 2: 115–30.
Fortun, Kim. 2001. *Advocacy after Bhopal: Environmentalism, Disaster, New Global Orders*. Chicago: University of Chicago Press.
Foucault, Michel. 1990 [1978]. *The History of Sexuality, Vol. 1: The Will to Knowledge*. Translated by Robert Hurley. London: Penguin.
———. 1990. *The History of Sexuality, Vol. 2: The Use of Pleasure*. Translated by Robert Hurley. New York: Vintage Books.
Fox, Warwick. 1995. "Transpersonal Ecology and the Varieties of Identification." In *The Deep Ecology Movement: An Introductory Anthology*, edited by A. Drengson and Y. Inoue, 136–54. Berkeley, CA: North Atlantic Books.
Franke, Richard W. 1993. *Life Is a Little Better: Redistribution as a Development Strategy in Nadur Village, Kerala*. Boulder, CO: Westview Press.
Franke, Richard W., and Barbara H. Chasin. 1992. *Kerala: Development through Radical Reform*. New Delhi: Promilla.
Freire, Paolo. 1970. *Pedagogy of the Oppressed*. New York: Herder and Herder.
———. 1974. *Education for Critical Consciousness*. London, UK: Sheed and Ward.
Freitag, Sandria B. 1989. *Collective Action and Community: Public Arenas and the Emergence of Communalism in North India*. Berkeley: University of California Press.

Gadgil, M., and R. Guha. 1994. "Ecological Conflicts and the Environmental Movement in India." *Development and Change* 25, no. 1: 101–36.
Gandhi, Mahatma, and Bharatan Kumarappa. 1954. *Nature Cure.*Ahmedabad: Navajivan Publishing House.
Gandhi, Mohandas K. 1928. *Satyagraha in South Africa.* Ahmedabad: Navajivan Publishing House.
———. 1954. *Autobiography: The Story of My Experiments with Truth.* Washington, DC: Public Affairs Press.
Gangadharan, M. *The Malabar Rebellion.* Kottayam, India: DC Books, 2008.
Gardner, Sarah Sturges. 1995. "Major Themes in the Study of Grassroots Environmentalism in Developing Countries." *Journal of Third World Studies* 12, no. 2: 200–44.
Gatt, C. 2017. *An Ethnography of Global Environmentalism: Becoming Friends of the Earth.* London: Routledge.
George, C. J. 2008. "Gōvindanṭe Mārgam." In *Putiya manuṣyan putiya lōkam: M. Gōvindanṭe cintakaḷ. By M. Govindan*, edited by C. J. George, 13–32. Kottayam, IN: DC Books
Gilio-Whitaker, Dina. 2019. *As Long as Grass Grows: The Indigenous Fight for Environmental Justice, from Colonization to Standing Rock.* Boston: Beacon Press.
Gilligan, C. 1993. *In a Different Voice: Psychological Theory and Women's Development.* Cambridge, MA: Harvard University Press.
Godwin, William. 1793. *Enquiry Concerning Political Justice and Its Influence on General Virtue and Happiness, Vol 1.* London: G. G. J. and J. Robinson.
Goffman, Erving. 1959. *The Presentation of Self in Everyday Life.* Garden City, NY: Doubleday.
———. 1974. *Frame Analysis: An Essay on the Organization of Experience.* New York: Harper & Row.
———. 1963. *Stigma: Notes on the Management of Spoiled Identity.* Englewood Cliffs, NJ: Prentice-Hall.
Goodpaster, Kenneth E. 1978. "On Being Morally Considerable." *The Journal of Philosophy* 75, no. 6: 308–25.
Gopikuttan, G. 1990. "House Construction Boom in Kerala: Impact on Economy and Society." *Economic and Political Weekly*, 25, no. 37: , 2083–88.
Gough, Kathleen. 1965a. "Village Politics in Kerala—I." *The Economic Weekly* 17, no. 8: 363–72.
———. 1965b. "Village Politics in Kerala—II." *The Economic Weekly* 17, no. 9: 413–20.
———. 1967. "Kerala Politics and the 1965 Elections." *International Journal of Comparative Sociology* 8, no. 1: 55–88.
———. 1968a. "Communist Rural Councillors in Kerala." *Journal of Asian and African Studies* 3: 181.
———. 1968b. "Peasant Resistance and Revolt in South India." *Pacific Affairs* 41, no. 4: 526–44.
Govindan, M. 2008a [1978]. "Janādhipatyam Socialism." In *Putiya manuṣyan putiya lōkam: M. Gōvindanṭe cintakaḷ*, edited by C. J. George, 518–23. Kottayam, IN: DC Books.

———. 2008b [1948]. “Talatiriñña Janakīyasāhityam.” In *Putiya manuṣyan putiya lōkam: M. Gōvindanṭe cintakaḷ*, edited by C. J. George, 764–77. Kottayam, IN: DC Books.

———. 2008c. *Putiya manuṣyan putiya lōkam: M. Gōvindanṭe cintakaḷ*, edited by C. J. George. Kottayam, IN: DC Books.

———. 2008d [1974]. *Navōtthanavum Kavitayum*. In *Putiya manuṣyan putiya lōkam: M. Gōvindanṭe cintakaḷ*, edited by C. J. George, 403-43. Kottayam, IN: DC Books.

Graeber, David. 2002. “The New Anarchists.” *New Left Review* 13, no. 6: 61–73.

———. 2009. *Direct Action: An Ethnography*. Oakland: AK Press.

Guha, Ranajit. 1982. “On Some Aspects of the Historiography of Colonial India.” In *Subaltern Studies I: Writings on South Asian History and Society*, edited by R. Guha, 1–8. Delhi: Oxford University Press.

Guha, Ramachandra. 1989. “Radical American Environmentalism and Wilderness Preservation: A Third World Critique.” *Environmental Ethics* 11, no. 1: 71–83.

———. 2000 [1989]. *The Unquiet Woods: Ecological Change and Peasant Resistance in the Himalaya*. Expanded ed. Berkeley: University of California Press.

———. 2013. “Mahatma Gandhi and the Environmental Movement in India.” In *Environmental Movements in Asia*, 65–82. New York: Routledge.

Guha, Ramachandra, and Joan Martinez-Alier. 2013. *Varieties of Environmentalism: Essays North and South*. London: Routledge.

Hacking, I. 1999. “Making Up People.” In *The Science Studies Reader*, edited by M. Biagioli, 161–71. New York: Routledge.

Habermas, Jürgen. 1989 [1962]. *The Structural Transformation of the Public Sphere: An Inquiry into a Category of Bourgeois Society*. Translated by Thomas Burger and Frederick Lawrence. Cambridge, MA: MIT Press.

Hale, Charles R. 2006. “Activist Research v. Cultural Critique: Indigenous Land Rights and the Contradictions of Politically Engaged Anthropology.” *Cultural Anthropology* 21, no. 1: 96–120.

Halliburton, Murphy. 2002. “Rethinking Anthropological Studies of the Body: Manas and Bōdham in Kerala.” *American Anthropologist* 104, no. 4: 1123–34.

Hanks, William. 1993. “Metalanguage and the Pragmatics of Deixis.” In *Reflexive Language: Reported Speech and Metapragmatics,* edited by John A. Lucy, 127–58. Cambridge, UK: Cambridge University Press.

Hansen, Thomas Blom. 1999. *The Saffron Wave: Democracy and Hindu Nationalism in Modern India*. Princeton, NJ: Princeton University Press.

Haraway, Donna. 1988. “Situated Knowledges: The Science Question in Feminism and the Privilege of Partial Perspective.” *Feminist Studies* 14, no. 3: 575–99.

Harikrishnan, S. 2023. *Social Spaces and the Public Sphere: A Spatial History of Modernity in Kerala*. London: Routledge.

———. 2020. “Communicating Communism: Social Spaces and the Creation of a ‘Progressive’ Public Sphere in Kerala, India.” *TripleC Special Issue: Communicative Socialism/Digital Socialism* 18, no. 1: 268–85.

Harrison, Faye V. 2010 [1997]. "Anthropology as an Agent of Transformation: Introductory Comments and Queries." In *Decolonizing Anthropology: Moving Further Toward an Anthropology for Liberation,* edited by Faye V. Harrison, 1–15. Arlington, VA: American Anthropological Association.

Harriss, John. 2011. "Civil Society and Politics: An Anthropological Perspective." In *A Companion to the Anthropology of India,* edited by Isabelle Clark-Deces, 389–406. Oxford: Wiley-Blackwell.

Heidegger, M. 1971. *Poetry, Language, Thought.* Translated by A. Hofstadter. New York: Harper and Row.

Heller, Patrick. 1999. *The Labor of Development: Workers and the Transformation of Capitalism in Kerala, India.* Ithaca, NY: Cornell University Press.

———. 2005. "Reinventing Public Power in the Age of Globalization: the Transformation of Movement Politics in Kerala." In *Social Movements in India: Poverty, Power, and Politics,* edited by Raka Ray and Mary Katzenstein, 79–106. Lanham, MD: Rowman and Littlefield.

Herman, Barbara. 1993. *The Practice of Moral Judgment.* Cambridge, MA: Harvard University Press.

Hofmeyr, Isabel. 2013. *Gandhi's Printing Press: Experiments in Slow Reading.* Cambridge, MA: Harvard University Press.

Holdrege, Barbara A. 1998. "Body Connections: Hindu Discourses of the Body and the Study of Religion." *International Journal of Hindu Studies* 2, no. 3: 341–86.

Howe, Cymene. *Intimate Activism: The Struggle for Sexual Rights in Postrevolutionary Nicaragua.* Durham, NC: Duke University Press, 2013.

Ingold, Tim. 1993. "Globes and Spheres: The Topology of Environmentalism." In *Environmentalism: The View from Anthropology,* edited by K. Milton, 31–42. New York: Routledge.

———. 2000. *The Perception of the Environment: Essays on Livelihood, Dwelling and Skill.* London: Routledge.

Isaac, T.M. Thomas, and Patrick Heller. 2003. "Democracy and Development: Decentralized Planning in Kerala." In *Deepening Democracy: Institutional Innovations in Empowered Participatory Governance*, edited by Archon Fung and Erik Olin Wright. London: Verso.

Isaac, T.M. Thomas, Richard Franke, and M.P. Parameswaran. 1997. "From Anti-Feudalism to Sustainable Development: The Kerala People's Science Movement." *Bulletin of Concerned Asian Scholars* 29: 34–44.

Isaac, T. M., and Richard W. Franke. 2002. *Local Democracy and Development: The Kerala People's Campaign for Decentralized Planning.* Lanham, MD: Rowman & Littlefield.

Jaffe, Alexandra M. 2009. "Introduction: The Sociolinguistics of Stance." In *Stance: Sociolinguistic Perspectives,* edited by Alexandra M. Jaffe, 3–28. Oxford: Oxford University Press.

Jaffry, Anwar, Mahesh Rangarajan, B. Ekbal, and K.P. Kannan. 1983. "Towards a People's Science Movement." *Economic And Political Weekly,* 18, no. 11:372–76.

Jain, Pankaj. 2011. *Dharma and Ecology of Hindu Communities: Sustenance and Sustainability*. London: Ashgate.

Jacob, J.C. 2010. *Johnsiyuṭe Ātmakatha*. Kozhikode, IN: Mathrubhumi Books.

Jaworska, Agnieszka. 2007. "Caring and Full Moral Standing." *Ethics* 117, no. 3: 460–97.

Jaworska, Agnieszka, and Julie Tannenbaum. 2021. "The Grounds of Moral Status." In *The Stanford Encyclopedia of Philosophy,* edited by Edward N. Zalta. https://plato.stanford.edu/archives/spr2021/entries/grounds-moral-status.

Jeffrey, Robin. 1976. "Temple-Entry Movement in Travancore, 1860–1940." *Social Scientist* 4, no. 8: 3–27.

———. 1978. "Matriliny, Marxism, and the Birth of the Communist Party in Kerala, 1930–1940." *The Journal of Asian Studies* 38, no. 1: 77–98.

———. 1993. *Politics, Women and Well Being: How Kerala Became "A Model."* New York: Oxford University Press.

———. 2009. "Testing Concepts About Print, Newspapers, and Politics: Kerala, India, 1800-2009." *The Journal of Asian Studies* 68, no. 2: 46589.

Johnson, Harriet McBryde. 2003. "Unspeakable Conversations." *New York Times Magazine,* February 16, 2003.

Juris, Jeffrey S. 2008. *Networking Futures: The Movements against Corporate Globalization.* Durham, NC: Duke University Press.

Kannan, Divya. 2012. "Socio-Religious Reform in Twentieth Century Kerala: Vagbhadananda and the Atma Vidya Sangham, 1900–40." *Proceedings of the Indian History Congress* 73: 1006–11.

Kannan, K. P. 2023. "Revisiting the Kerala 'Model' of Development: A Sixty-Year Assessment of Successes and Failures." *The Indian Economic Journal* 71, no. 1: 120–51.

Kant, Immanuel. 1998 [1785]. *Groundwork of the Metaphysics of Morals.* Translated by Mary J. Gregor. Cambridge, UK: Cambridge University Press.

Karlsson, Bengt G., and Tanka B. Subba. 2006. *Indigeneity in India.* London, UK: Kegan Paul.

Kavedžija, Iza. 2016. *Values of Happiness: Toward an Anthropology of Purpose in Life.* Chicago: Hau Books.

Keane, Webb. 1997. *Signs of Recognition: Powers and Hazards of Representation in an Indonesian Society.* Berkeley: University of California Press.

———. 2008. "The Evidence of the Senses and the Materiality of Religion." *Journal of the Royal Anthropological Institute* 14: S110-27.

———. 2016. *Ethical Life: Its Natural and Social Histories.* Princeton: Princeton University Press.

King, Martin Luther, Jr. 1997 [1963]. "Letter from Birmingham jail." In *The Norton Anthology of African American Literature,* edited by Henry Louis Gates, Jr., and Nellie Y. McKay. New York: W.W. Norton.

Kirksey, S. Eben, and Stefan Helmreich. 2010. "The Emergence of Multispecies Ethnography." *Cultural Anthropology* 25, no. 4: 545–76.

Kleinman, Arthur. 2009. "Caregiving: The Odyssey of Becoming More Human." *The Lancet* 373: 292–93.

Klein, Jakob A. 2008. "Afterword: Comparing Vegetarianisms." *South Asia: Journal of South Asian Studies* 31, no. 1: 199–212.

Kopnina, Helen. 2014. "Environmental Justice and Biospheric Egalitarianism: Reflecting on a Normative-Philosophical View of Human-Nature Relationship." *Earth Perspectives* 1, no. 1: 1–11.

———. 2016. "Of Big Hegemonies and Little Tigers: Ecocentrism and Environmental Justice." *The Journal of Environmental Education* 47, no. 2: 139–50.

Krings, Amy, Bryan G. Victor, John Mathias, and Brian E. Perron. 2018. "Environmental Social Work in the Disciplinary Literature, 1991–2015." *International Social Work* 63, no. 3: 275–90.

Kunjan Pillai, Suranad, ed. 1965. *Malayalam Lexicon*. Vol. 1. Thiruvananthapuram, IN: University of Kerala.

Laclau, Ernesto. 2005. *On Populist Reason*. New York: Verso.

Laidlaw, James. 1995. *Riches and Renunciation: Religion, Economy, and Society among the Jains*. Oxford: Oxford University Press.

———. 2002. "For an Anthropology of Ethics and Freedom." *The Journal of the Royal Anthropological Institute* 8, no. 2: 311–32.

———. 2005. "A Life Worth Leaving: Fasting to Death as Telos of a Jain Religious Life." *Economy and Society* 34, no. 2: 178–99.

———. 2010. "Ethical Traditions in Question: Diaspora Jainism and the Environmental and Animal Liberation Movements." In *Ethical Life in South Asia*, edited by Anand Pandian and Daud Ali, 61–82. Bloomington: Indiana University Press.

———. 2014. *The Subject of Virtue: An Anthropology of Ethics and Freedom*. Cambridge, UK: Cambridge University Press.

Lal, Vinay. 2000. "Gandhi and the Ecological Vision of Life." *Environmental Ethics* 22, no. 2: 149–68.

Lakier, Genevieve. 2014. *The Spectacle of Power: Coercive Protest and the Problem of Democracy in Nepal*. Chicago: The University of Chicago Press.

Lambek, Michael. 2010. "Introduction." In *Ordinary Ethics: Anthropology, Language, and Action*, edited by M. Lambek, 1–38. New York: Fordham University Press.

———. 2011. "Catching the Local." *Anthropological Theory* 11, no. 2: 197–221.

———. 2015. "On the Immanence of the Ethical: A Response to Michael Lempert, 'No ordinary ethics'." *Anthropological Theory* 15, no. 2: 128–32.

Latour, Bruno. 2000. "When Things Strike Back: A Possible Contribution of 'Science Studies' to the Social Sciences." *The British Journal of Sociology* 51, no. 1: 107–23.

———.2005. *Reassembling the Social: An Introduction to Actor-Network Theory*. Oxford: Oxford University Press.

Lemercinier, Geneviève. 1984. *Religion and Ideology in Kerala*. New Delhi: D.K. Agencies.

Lempert, Michael. 2008. "The Poetics of Stance: Text-Metricality, Epistemicity, Interaction." *Language in Society* 37, no. 4: 569–92.

———. 2013. "No Ordinary Ethics." *Anthropological Theory* 13, no. 4: 370–93.

———. 2015. "Ethics Without Immanence: A Reply to Michael Lambek." *Anthropological Theory* 15, no. 2: 133–40.

Leopold, Aldo. 1949. *A Sand County Almanac*. New York: Oxford University Press.

———. 1966 [1953]. "Round River." In *A Sand County Almanac with Other Essays on Conservation from Round River*. New York: Oxford University Press.

Lindberg, Anna. 2005. *Modernization and Effeminization in India: Kerala Cashew Workers since 1930*. Copenhagen: NIAS.

Lukose, Ritty. 2009. *Liberalization's Children: Gender, Youth, and Consumer Citizenship in Globalizing India*. Durham, NC: Duke University Press.

MacKinnon, Catharine A. 1982. "Feminism, Marxism, Method, and the State: An Agenda for Theory." *Signs: Journal of Women in Culture and Society* 7, no. 3: 515–44.

MacIntyre, Alasdair. 1966. *A Short History of Ethics*. New York: Macmillan.

———. 1981. *After Virtue: A Study in Moral Theory*. Notre Dame, IN: University of Notre Dame Press.

Mackenzie, John S., and Martyn Jeggo. 2019. "The One Health Approach—Why Is It So Important?" *Tropical Medicine and Infectious Disease* 4, no. 2: 88.

Mahmood, Saba. 2005. *Politics of Piety: The Islamic Revival and the Feminist Subject*. Princeton: Princeton University Press.

Manjusha, K. 2016. "Silent Valley Movement in Kerala: A Study on the Contributions of Kerala Sastra Sahitya Parishad." *International Journal of Research in Social Sciences* 6, no. 3: 267–74.

Mannathukkaren, Nissim. 2016. "Communalism Sans Violence: A Keralan Exceptionalism?" *Sikh Formations* 12, no. 2–3: 223–42.

Marriott, McKim. 1968. "Caste Ranking and Food Transactions: A Matrix Analysis." In *Structure and Change in Indian Society,* edited by Milton B. Singer and Bernard S. Cohn, 133–72. Chicago: Aldine Publishing Co.

Marriott, McKim, and Ronald B. Inden. 2011. "Toward an Ethnosociology of South Asian Caste Systems." In *The New Wind,* edited by Kenneth David, 227–38. Berlin, New York: De Gruyter Mouton.

Martellozzo, Nicola. 2021. "Wind, Wood, and the Entangled Life of Disasters." *HAU: Journal of Ethnographic Theory* 11, no. 2: 428–44.

Martin, Christian, ed. 2018. *Language, Form(s) of Life, and Logic*. Berlin: De Gruyter.

Martinez-Alier, Joan. 2002. *The Environmentalism of the Poor: A Study of Ecological Conflicts and Valuation*. Cheltenham, UK: Edward Elgar Publishing.

———. 2016. "Global Environmental Justice and the Environmentalism of the Poor." In *the Oxford Handbook of Environmental Political Theory,* edited by Teena Gabrielson, Cheryl Hall, John M. Meyer and David Schlosberg, 547–62. Oxford: Oxford University Press.

Martinez-Alier, Joan, Leah Temper, Daniela Del Bene, and Arnim Scheidel. 2016. "Is There a Global Environmental Justice Movement?" *The Journal of Peasant Studies* 43, no. 3: 731–55.

Mason, Lisa Reyes, and Jonathan Rigg. 2019. *People and Climate Change: Vulnerability, Adaptation, and Social Justice*. Oxford: Oxford University Press.

Mathew, Leya. 2022. *English Linguistic Imperialism from Below: Moral Aspiration and Social Mobility*. Bristol, UK: Multilingual Matters.

Mathews, Freya. 2013. "Moral Ambiguities in the Politics of Climate Change." In *Climate Change and Environmental Ethics*, edited by Ved P. Nanda, 43–64. New Brunswick, NJ: Transaction Publishers.

Mathias, John. 2010. "Of Contract and Camaraderie: Thoughts on What Relationships in the Field Could Be." *Collaborative Anthropologies* 3, no. 1: 110–20.

———. 2019. "Sticky Ethics: Environmental Activism and the Limits of Ethical Freedom in Kerala, India." *Anthropological Theory* 20, no. 3: 253–76.

———. 2023. "Which Environmental Social Work? Environmentalisms, Social Justice, and the Dilemmas Ahead." *Social Service Review* 97, no. 3: 569–601.

Mattingly, Cheryl. 2012. "Two Virtue Ethics and the Anthropology of Morality." *Anthropological Theory* 12, no. 2: 161–84.

———. 2014. *Moral Laboratories: Family Peril and the Struggle for a Good Life*. Berkeley: University of California Press.

Mattingly, Cheryl, and Jason Throop. 2018. "The Anthropology of Ethics and Morality." *Annual Review of Anthropology* 47: 475–92.

Mazzarella, William. 2006. "Internet X-ray: E-governance, Transparency, and the Politics of Immediation in India." *Public Culture* 18, no. 3: 473–505.

McGarr, Paul Michael. 2014. "'Quiet Americans in India': The CIA and the Politics of Intelligence in Cold War South Asia*1." *Diplomatic History* 38, no. 5: 1046–82.

McGurty, Eileen. 2009. *Transforming Environmentalism: Warren County, PCBs, and the Origins of Environmental Justice*. New Brunswick: Rutgers University Press.

McKee, Emily. 2015. "Trash Talk: Interpreting Morality and Disorder in Negev/Naqab Landscapes." *Current Anthropology* 56, no. 5: 733–52.

Mead, George Herbert. 1934. *Mind, Self & Society from the Standpoint of a Social Behaviorist*. Chicago: University of Chicago Press.

Melosi, Martin. V. 2000. "Environmental Justice, Political Agenda Setting, and the Myths of History." *Journal of Policy History* 12, no. 1: 43–71.

Méndez, Michael. 2020. *Climate Change from the Streets*. New Haven: Yale University Press.

Menon, Dilip. 1994. *Caste, Nationalism, and Communism in South India: Malabar, 1900–1948*. Cambridge, UK: Cambridge University Press.

Menon, Dilip, and Sasikumar Harikrishnan. 2023. "Kozhikode's Kōlāya Gatherings: Conversations That Shaped Kerala Modernity in '60s." *The News Minute*. https://www.thenewsminute.com/kerala/kozhikode-s-k-l-ya-gatherings-conversations-shaped-kerala-modernity-60s-177066.

Meyer, Birgit. 2011. "Mediation and Immediacy: Sensational Forms, Semiotic Ideologies and the Question of the Medium." *Social Anthropology* 19, no. 1: 23–39.

Mines, Mattison. 1994. *Public Faces, Private Voices: Community and Individuality in South India*. Berkeley: University of California Press.

Mitchell, Lisa. 2023. *Hailing the State*. Durham: Duke University Press.

Mithun, S. 2012. "Preferences of 24 Hours News Channel Viewers in Kerala." PhD diss., Christ University Bangalore.

Moag, Rodney. 1980. *Malayalam: A University Course and Reference Grammar*. Ann Arbor: University of Michigan.

Mol, Annemarie. 2008. *The Logic of Care: Health and the Problem of Patient Choice*. London: Routledge.

Morgan, E.S. 1989. *Inventing the People: The Rise of Popular Sovereignty in England and America*. New York: Norton.

Morton, Gregory. Forthcoming. *Return from the World: Economic Growth and Reverse Migration in Brazil*. Chicago: University of Chicago Press.

Moyal-Sharrock, Danièle. 2015. "Wittgenstein on Forms of Life, Patterns of Life, and Ways of Living." *Nordic Wittgenstein Review*: 21–42.

Næss, Arne. 1973. "The Shallow and the Deep, Long-Range Ecology Movement: A Summary." *Inquiry* 16, no. 1–4: 95–100.

———. 2008. "Self-Realization: An Ecological Approach to Being in the World." In *The Ecology of Wisdom: Writings by Arne Næss,* edited by A. Næss, A. Drengson, and B. Devall, 81–98. Berkeley, CA: Counterpoint.

Namboodiri, D. Damodaran. 1999. "Caste and Social Change in Colonial Kerala." In *Perspectives on Kerala History: The Second Millenium,* edited by P.J. Cherian, 426–55. Thiruvananthapuram, IN: Kerala Council for Historical Research.

Namboodiripad, E.M.S. 1976. *How I Became a Communist*. Trivandrum: Chinta Publishers.

Nandy, Ashis. 1983. *The Intimate Enemy: Loss and Recovery of Self under Colonialism*. Delhi: Oxford University Press.

Nash, Roderick Frazier. 1989. *The Rights of Nature: A History of Environmental Ethics*. Madison: University of Wisconsin Press.

Nisar, M. and Meena Kandasamy. 2007. *Ayyankali: A Dalit Leader of Organic Protest*. Calicut, Kerala: Other Books.

Nossiter, T.J. 1982. *Communism in Kerala: A Study in Political Adaptation*. Delhi, India: Oxford University Press.

Ormrod, James S. 2011. "'Making Room for the Tigers and the Polar Bears': Biography, Phantasy and Ideology in the Voluntary Human Extinction Movement." *Psychoanalysis, Culture & Society* 16, no. 2: 142–61. doi: 10.1057/pcs.2009.30.

Ortner, S.B. 1984. "Theory in Anthropology since the Sixties." *Comparative Studies in Society and History* 26, no. 1: 126–66.

———. 2016. "Dark Anthropology and Its Others: Theory Since the Eighties." *HAU: Journal of Ethnographic Theory* 6, no. 1: 47–73.

Osella, Filippo, and Caroline Osella. 2000. *Social Mobility in Kerala: Modernity and Identity in Conflict*. Sterling, VA: Pluto Press.

Osella, Caroline, and Filippo Osella. 2008. "Food, Memory, Community: Kerala as Both 'Indian Ocean' Zone and as Agricultural Homeland." *South Asia: Journal of South Asian Studies* 31, no. 1: 170–98.

Ottinger, Gwen, and Benjamin R. Cohen. 2011. *Technoscience and Environmental Justice Expert Cultures in a Grassroots Movement*. Cambridge, MA: MIT Press.

Pandey, Gyanendra. 2006. *The Construction of Communalism in Colonial North India*. Delhi: Oxford University Press.

Pandian, Anand. 2009. *Crooked stalks: Cultivating virtue in South India*. Durham. NC: Duke University Press.

Panangad, P. 2018. *Malayāla samāntara māsika caritram*. Thiruvananthapuram: State Institute of Languages.

Panikkar, Kandiyur Narayana. 1989. *Against Lord and State: Religion and Peasant Uprisings in Malabar, 1836–1921*. Oxford: Oxford University Press.

Pellow, David Naguib. 2017. *What is Critical Environmental Justice?* New York: John Wiley & Sons.

Pellow, David N., and Robert J. Brulle. 2005. *Power, Justice, and the Environment: A Critical Appraisal of the Environmental Justice Movement*. Cambridge, MA: MIT Press.

PETA. n.d. "What is Speciesism?" People for the Ethical Treatment of Animals (PETA). Accessed November 16, 2021. https://www.peta.org/features/what-is-speciesism.

Pezzullo, Phaedra C. 2009. *Toxic Tourism: Rhetorics of Pollution, Travel, and Environmental Justice*. Tuscaloosa: University of Alabama Press.

Pezzullo, Phaedra C., and Ronald D. Sandler. 2007. *Environmental Justice and Environmentalism: The Social Justice Challenge to the Environmental Movement*. Cambridge, MA: MIT Press.

Plumwood, Val. 1993. *Feminism and the Mastery of Nature*. London: Routledge.

Polletta, Francesca. 1997. "Culture and Its Discontents: Recent Theorizing on the Cultural Dimensions of Protest." *Sociological Inquiry* 67, no. 4: 431–50.

———. 2006. *It Was like a Fever: Storytelling in Protest and Politics*. Chicago: University of Chicago Press.

———. 2008. "Culture and Movements." *The Annals of the American Academy of Political and Social Science* 619, no. 1: 78–96.

———. 2012. *Freedom is an Endless Meeting*. Chicago: University of Chicago Press.

Pomerantz, Anita. 1984. "Agreeing and Disagreeing with Assessments: Some Features of Preferred/Dispreferred Turn Shapes." In *Structures of Social Action: Studies in Conversation Analysis*, edited by J. M. Atkinson and J. Heritage, 55–101. Cambridge, UK: Cambridge University Press.

Powell, Dana E. 2020. "Comments on Anthropology of Activism." In *Anthropology and Activism*, edited by A.J. Willow and K.A. Yotebieng, 85–97. London: Routledge.

Prakash, B. A. 2004. "Economic Backwardness and Economic Reforms in Kerala." In *Kerala's Economic Development: Performance and Problems in the Post-Liberalization Period*, edited by B. A. Prakash, 32–60. New Delhi: Sage.

Prasse-Freeman, Elliott. 2023. *Rights Refused: Grassroots Activism and State Violence in Myanmar*. Stanford, CA: Stanford University Press.

Pulido, Laura. 1996. *Environmentalism and Economic Justice: Two Chicano Struggles in the Southwest*. Tucson: University of Arizona Press.

Radhakrishnan, A. 2008. *M. Govindan: Jīvitavum Āśayavum*. Kottayam, IN: DC Books.

Radhakrishnan, P. 1981. "Land Reforms in Theory and Practice: The Kerala Experience." *Economic And Political Weekly* 16, no. 52: A129–A137.

Ramachandran, V.K. 2000. "Human Development Achievements in an Indian State: A Case Study of Kerala." In *Social Development and Public Policy: A Study of Some Successful Experiences*, edited by Dharam Ghai, 46–102. London: Palgrave Macmillan UK.

Rammohan, K.T. 2008. "Caste and Landlessness in Kerala: Signals from Chengara." *Economic and Political Weekly* 43 , no. 37: 14–16.

———. 2010. "Caste, Public Action and the Kerala Model." In *Development, Democracy and the State*, edited by K. Raviraman, 35–49. London: Routledge.

Rancière, Jacques. 1999. *Dis-agreement: Politics and Philosophy*. Translated by Julie Rose. Minneapolis: University of Minnesota Press.

Reno, Joshua. 2011. "Beyond Risk: Emplacement and the Production of Environmental Evidence." *American Ethnologist* 38, no. 3: 516–30.

Robbins, J. 2004. *Becoming Sinners: Christianity and Moral Torment in a Papua New Guinea Society*. Berkeley: University of California.

———. 2007. "Between Reproduction and Freedom: Morality, Value, and Radical Cultural Change." *Ethnos* 72, no. 3: 293–314.

———. 2016. "What Is the Matter with Transcendence? On the Place of Religion in the New Anthropology of Ethics." *Journal of the Royal Anthropological Institute* 22, no. 4: 767–81.

Rootes, Christopher. 2007. "Acting Locally: The Character, Contexts and Significance of Local Environmental Mobilisations." *Environmental Politics* 16, no. 5: 722–41.

Roy, Arundhati. 1997. *The God of Small Things*. New York: Random House.

———. 1999. "The Greater Common Good." In *The Cost of Living*, edited by A. Roy, 1–102. London: Flamingo.

Roy, M.N. 1952. *Radical Humanism*. New Delhi: Janta Press.

Roy, M.N., and S. Philip. 1968. *Beyond Communism*. Kolkata: Renaissance Publishers.

Roy, Srirupa. 2007. *Beyond Belief: India and the Politics of Postcolonial Nationalism*. Durham: Duke University Press.

Ryder, Richard D. 2010. "Speciesism Again: The Original Leaflet." *Critical Society* 2, no. 1: 2.

Sahlins, Marshall. 2017 [1972]. *Stone Age Economics*. New York: Routledge.

Santha, E.K. 2016. "Marxist Praxis: Communist Experience in Kerala: 1957–2011." PhD diss., Sikkim University.

Sartre, Jean-Paul. 2007 [1946]. *Existentialism is a Humanism*. Translated by Carol Macomber. New Haven: Yale University Press.

Schaler, Jeffrey A., ed. 2009. *Peter Singer under Fire: The Moral Iconoclast Faces His Critics*. Peru, IL: Carus Publishing.

Scheper-Hughes, Nancy. 1995. "The Primacy of the Ethical: Propositions for a Militant Anthropology." *Current Anthropology* 36, no. 3: 409–40.

Schlosberg, David. 2007. *Defining Environmental Justice: Theories, Movements, and Nature*. Oxford: Oxford University Press.

———. 2013. "Theorising Environmental Justice: The Expanding Sphere of a Discourse." *Environmental Politics* 22, no. 1: 37–55.

Schutz, Alfred. 1967. *The Phenomenology of the Social World.* Evanston, IL: Northwestern University Press.

Schwenkler, John. 2017. Review of *Puzzling Identities*, Vincent Descombes, Stephen Adam Schwartz. *Mind* 126: 967–74.

Sen, Amartya. 2000. *Amartya Sen on Kerala.* New Delhi: Institute of Social Sciences.

Sessions, G. 1995. "Ecocentrism and the Anthropocentric Detour." In *Deep Ecology for the 21st Century,* edited by G. Sessions, 156–84. Boston, MA: Shambhala.

Shah, Alpa. 2010. *In the Shadows of the State: Indigenous Politics, Environmentalism, and Insurgency in Jharkhand, India.* Durham, NC: Duke University Press.

———. 2022. "Why I Write? In a Climate against Intellectual Dissidence." *Current Anthropology* 63, no. 5: 570-=-600. doi: 10.1086/722030.

Shaji, K. A. 2020. "India's First Environmental Magazine Completes 40 Years of Effective Interventions." *The Citizen.* https://www.thecitizen.in/index.php/en/newsdetail/index/13/19188/indias-first-environmental-magazine-completes-40-years-of-effective-interventions.

Shapiro, Nicholas. 2015. "Attuning to the Chemosphere: Domestic Formaldehyde, Bodily Reasoning, and the Chemical Sublime." *Cultural Anthropology* 30, no. 3: 368–93.

Sheldon, Victoria. 2020. "Vitality, Self-healing and Ecology: The Flow of Naturopathic Thought Across the United States and India." *Society and Culture in South Asia* 6, no. 1: 121–43.

Shrijan, V. C. 2012. "'Samāntaravum' 'mukhyadhāravum' tammil ā paḻaya yuddham ippōḻilla." *Mathrubhumi* 90 42: 8–12.

Sidnell, Jack. 2010. *Conversation Analysis: An Introduction.* Chichester, UK: Wiley-Blackwell.

———. 2017. "Askesis and the Ethics of Eating at a Yoga School in Southern India." In *Gastronomy, Culture, and the Arts: A Scholarly Exchange of Epic Portions,* edited by T. Lobalsamo, A. Pasquali and C. Lebrec, 159–74. Toronto: Legas Publishing.

Sidnell, Jack, Marie Meudec, and Michael Lambek. 2019. *Ethical Immanence.* London: Sage.

Silverstein, Michael. 1976. "Shifters, Linguistic Categories, and Cultural Description." In *Meaning in Anthropology,* edited by Keith H. Basso and Henry A. Selby, 11–55. Albuquerque: University of New Mexico Press.

Singer, Peter. 1979. *Practical Ethics.* Cambridge: Cambridge University Press.

———. 1981. *The Expanding Circle: Ethics and Sociobiology.* New York: Farrar, Straus & Giroux.

———. 2009. "Speciesism and Moral Status." *Metaphilosophy* 40, no. 3–4: 567–581. https://doi.org/10.1111/j.1467–9973.2009.01608.x.

Singh, Prerna. 2011. "We-ness and Welfare: A Longitudinal Analysis of Social Development in Kerala, India." *World Development* 39, no. 2: 282–93.

———. 2015. *How Solidarity Works for Welfare: Subnationalism and Social Development in India.* Cambridge, UK: Cambridge University Press.

Sitrin, Marina. 2006. *Horizontalism: Voices of Popular Power in Argentina.* Oakland, CA: AK Press.

Skaria, Ajay. 2010. "Living by Dying: Gandhi, *Satyagraha,* and the Warrior." In *Ethical Life in South Asia,* edited by Anand Pandian and Daud Ali, 211–31. Bloomington: Indiana University Press.

Smith, Adam. 2002 [1761]. *The Theory of Moral Sentiments.* Cambridge: Cambridge University Press.

Smith, Kimberly K. 2021. *African American Environmental Thought: Foundations.* Lawrence: University Press of Kansas.

Snow, D.A., and R.D. Benford. 1988. "Ideology, Frame Resonance, and Participant Mobilization." *International Social Movement Research* 1, no. 1: 197–217.

Speed, Shannon. 2006. "At the Crossroads of Human Rights and Anthropology: Toward a Critically Engaged Activist Research." *American Anthropologist* 108, no. 1: 66–76.

Spivak, Gayatri Chakravorty. 1988. "Can the Subaltern Speak?" In *Marxism and the Interpretation of Culture,* edited by C. Nelson and L. Grossberg, 271–313. Urbana: University of Illinois Press

S.P., John. 2012. "History of the Left Intervention in the Cultural Scenario of Kerala." PhD dissertation, University of Kerala.

Sreejith, K. 2005. "Naxalite Movement and Cultural Resistance: Experience of Janakiya Samskarika Vedi in Kerala (1980–82)." *Economic And Political Weekly,* 40, no. 50: 5333–37.

Sreekumar, T.T., and Govinda Parayil. 2006. "Interrogating Development: New Social Movements, Democracy, and Indigenous People's Struggles in Kerala." In *Kerala: The Paradoxes of Public Action and Development,* edited by Joseph Tharamangalam, 217–57. New Delhi: Orient Longman.

———. 2010. "Social Space, Civil Society and Transformative Politics of New Social Movements in Kerala." In *Development, Democracy and the State: Critiquing the Kerala Model of Development,* edited by K. Ravi Raman, 237–53. New York, NY: Routledge.

Sreemahadevan Pillai, P.R. 2008. *The Saga of Plachimada.* Mumbai: Vikas Adhyayan Kendra.

Srinivas, Mysore Narasimhachar. 1966. *Social Change in Modern India.* Berkeley: University of California Press.

Stasch, Rupert. 2009. *Society of Others: Kinship and Mourning in a West Papuan Place.* Berkeley: University of California Press.

Steur, Luisa. 2010. "Adivasi Workers' Struggles and the Kerala Model: Interpreting the Past, Confronting the Present." In *Development, Democracy and the State,* edited By K. Ravi Raman, 231–46. London: Routledge.

———. 2011. "Adivasis, Communists, and the Rise of Indigenism in Kerala." *Dialectical Anthropology* no. 35 (1):59-76.

———. 2017. *Indigenist Mobilization: Confronting Electoral Communism and Precarious Livelihoods in Post-Reform Kerala.* New York: Berghahn Books.

Sunilraj, Balu. 2023. "The Local Roots of Communist Support in Kerala." Commonwealth & Comparative Politics 61, no. 1: 65–89.

Sze, Julie, and Jonathan K. London. 2008. "Environmental Justice at the Crossroads." *Sociology Compass* 2, no. 4: 1331–54. https://doi.org/10.1111/j.1751-9020.2008.00131.x.

Taylor, Charles. 1976. "Responsibility for Self." In *The Identities of Persons*, edited by Amelie Oksenberg Rorty, 281–99. Berkeley: University of California Press.

Taylor, Dorceta E. 1993. "Environmentalism and the Politics of Inclusion." In *Confronting Environmental Racism: Voices from the Grassroots,* edited by Robert D. Bullard, 53–62. Boston, MA: South End Press.

———. 2000. "The Rise of the Environmental Justice Paradigm: Injustice Framing and the Social Construction of Environmental Discourses." *American Behavioral Scientist* 43, no. 4: 508–80.

———. 2014. *Toxic Communities.* New York: New York University Press.

Temper, Leah, Daniela Del Bene, and Joan Martinez-Alier. 2015. "Mapping the Frontiers and Front Lines of Global Environmental Justice: The EJAtlas." *Journal of Political Ecology* 22, no. 1: 255–78.

Tesh, Sylvia. 2000. *Uncertain Hazards: Environmental Activists and Scientific Proof.* Ithaca, NY: Cornell University Press.

Tharamangalam, Joseph. 1998. "The Perils of Social Development without Economic Growth: The Development Debacle of Kerala, India." *Bulletin of Concerned Asian Scholars* 30, no. 1: 23–34.

———. 2007. "Why Pursuing the Public Good Matters: Lessons from Kerala and Cuba." In *A Decade of Decentralization in Kerala: Experience and Lessons,* edited by M.A. Oommen, 119–40. New Delhi: Institute of Social Sciences.

Thiranagama, Sharika. 2019. "Respect Your Neighbor as Yourself: Neighborliness, Caste, and Community in South India." *Comparative Studies in Society and History* 61, no. 2: 269–300.

Thomas, Sonja. 2018. *Privileged Minorities: Syrian Christianity, Gender, and Minority Rights in Postcolonial India.* Seattle: University of Washington Press.

Thomas, Vinod. "Kerala: A Paradox or Incomplete Agenda?" In *Kerala: The Paradoxes of Public Action and Development*, edited by Joseph Tharamangalam, 69–93. New Delhi: Orient Longman, 2006.

Throop, C.J. 2010. *Suffering and Sentiment: Exploring the Vicissitudes of Experience and Pain in Yap*. Berkeley: University of California Press.

Ticktin, Miriam I. 2019. "From the Human to the Planetary." *Medicine Anthropology Theory* 6, no. 3: 133–60.

Trisal, Nishita. n.d. "How to Sustain a Strike: Rules, Routines, and the Essential in Kashmir" (unpublished paper).

Tronto, Joan C. 1993. *Moral Boundaries: A Political Argument for an Ethic of Care.* London, UK: Psychology Press.

Trouillot, Michel-Rolph. 2003. *Global Transformations: Anthropology and the Modern World.* New York: Palgrave MacMillan.

Tsing, Anna Lowenhaupt. 2000. "The Global Situation." *Cultural Anthropology* 15, no. 3: 327–60.

———. 2005. *Friction: An Ethnography of Global Connection*. Princeton, NJ: Princeton University Press.

Vaughn, Sarah E., and Daniel Fisher. 2021. "Witnessing Environments." *HAU: Journal of Ethnographic Theory* 11, no. 2: 387–94. doi: 10.1086/716548.

Venkat, Bharat Jayram. 2021. *At the Limits of Cure*. Durham, NC: Duke University Press.

Venkatesan, Soumhya. 2021. "The Wedding of Two Trees: Connections, Equivalences, and Subjunctivity in a Tamil Ritual." *Journal of the Royal Anthropological Institute* 27, no. 3: 478–95.

———. 2023. "Freedom." In *The Cambridge Handbook for the Anthropology of Ethics*, edited by James Laidlaw, 251–80. Cambridge, UK: Cambridge University Press.

Vijayalekshmy, M. 1999. "'Pattini'—An Institution of Fasting Observed for the Redressal of Grievances in Pre-Colonial Kerala." *Proceedings of the Indian History Congress* 60: 413–16.

Warren, Mary Anne. 1997. *Moral Status: Obligations to Persons and Other Living Things*. Oxford: Clarendon Press.

Washington, Haydn, Guillaume Chapron, Helen Kopnina, Patrick Curry, Joe Gray, and John J. Piccolo. 2018. "Foregrounding Ecojustice in Conservation." *Biological Conservation* 228: 367–74.

Weber, Max. 1968 [1922]. *Economy and Society: An Outline of Interpretive Sociology*. Berkeley: University of California Press.

White, Lynn. 1967. "The Historical Roots of Our Ecologic Crisis." *Science* 155, no. 3767: 1203–07.

Whyte, Kyle. 2015. "How Similar Are Indigenous North American and Leopoldian Environmental Ethics?" Available at SSRN 2022038.

Whyte, Kyle Powys, and Chris J Cuomo. 2016. "Ethics of Caring in Environmental Ethics: Indigenous and Feminist Philosophies." In *The Oxford Handbook of Environmental Ethics*, edited by Stephen M. Gardiner and Allen Thompson, 234–47. New York: Oxford University Press.

Williams, Bernard. 1981. *Moral Luck: Philosophical Papers 1973–1980*. Cambridge: Cambridge University Press.

———. 1985. *Ethics and the Limits of Philosophy*. Cambridge, MA: Harvard University Press.

———. 2009. "The Human Prejudice." In *Peter Singer Under Fire: The Moral Iconoclast Faces His Critics*, edited by Jeffrey A. Schaler, 77–96. Chicago: Carus Publishing Company.

Wolf, David B. 2000. "Social Work and Speciesism." *Social Work* 45, no. 1: 88–93.

Wright, Fiona. 2016. "Palestine, My Love: The Ethico-Politics of Love and Mourning in Jewish Israeli Solidarity Activism." *American Ethnologist* 43, no. 1: 130–43.

Yates, Luke. 2015. Rethinking Prefiguration: Alternatives, Micropolitics and Goals in Social Movements. *Social Movement Studies* 14, no. 1: 1–21.

Zigon, Jarrett. 2008. *Morality: An Anthropological Perspective*. Oxford: Berg.

———. 2014. "An Ethics of Dwelling and a Politics of World-Building: A Critical Response to Ordinary Ethics." *Journal of the Royal Anthropological Institute* 20, no. 4: 746–64.

———. 2021. "How Is It Between Us? Relational Ethics and Transcendence." *Journal of the Royal Anthropological Institute* 27, no. 2: 384–401.

Zuckerman, Charles H. P. 2018. "Good Gambling: Meaning and Moral Economy in Late-Socialist Laos." PhD diss., University of Michigan.

———. 2023. "Video Footage and the Grain of Practice." *HAU: Journal of Ethnographic Theory* 13, no. 1: 128–45. doi: 10.1086/725027.

# Index


Achuthanandan, V.S., 145–46, 159
action councils, 32, 49, 158. *See also* Manamur Action Council
activism: anthropology and, 12–13, 45–46, 194n33, 201n83; community affiliation and, 7–11, 73–76; influencing others and, 12, 123–25, 136, 170; as life choice, 3–6, 14–15, 34–37, 165–66, 192n24, 225n5; norm-breaking and, 5–6, 37, 58–59, 194n32; purposes and, 11–12, 31, 42, 61, 64, 168, 194n32; science and, 137–143, 195n35; two approaches, 2–3, 14–15, 23–24, 97. *See also* activist ethics; environmental justice; local activists; protest; radical activists
activist ethics: defined by Faiza, 5–6, 165–67; divisiveness of, 5–6, 73, 101, 110–13, 124–25, 217n22; as exemplary, 5, 168, 194, 218n26; impartiality and, 83, 88–89, 165–69, 220n14, 224n2; limits of, 59–61, 122–25, 158–59, 177–79; as meta-ethic, 106, 107–8, 121, 216n7
Adarsh: editorial work, 40–41, 46–49, 51–52, 54; in movie theater, 101–3, 123; opposition to family, 118–122, 123–24; tea drinking, 118; VapoRub, 117–18
*adivasis,* 97, 191, 214n41
agency, 12, 129, 156, 193, 191n19, 206n34
Ahmann, Chloe, 209n11
Ahmed: as activist, 2, 25–27, 58, 163–65, 176–78; career, 26, 178; Dialogue Journey dilemma, 1–3, 25, 172–73, 177–78; farewell party, 161–62; Muslim faith, 23, 25; plastic bags story, 162–63; at quarry, 173–75; research assistant, 2, 25–26, 182–84
allopathic medicine. *See* English medicine
Alter, Joseph, 217n18, 218n25
alternative medicine. *See* English medicine
alternatives: alienation and, 104, 117, 122–25, 217n22; change and, 110–14; difference and, 36–37, 100–1, 104, 208; education and, 53, 225; ethical consumption and, 112; *jāppi,* 104, 110, 113–14; Leftism and, 37–39, 44–45, 204n19; levels of, 104; mainstream and, 55, 103, 179, 192n20, 201n1, 206n30, 208n44; Malayalam term, 37, 103, 203n10; media and, 37, 51–52; politics of, 112–14, 120, 123, 203n10, 217n16. *See also* little magazines; nature cure
anthropocentrism, 6, 9–10, 83, 188n7, 191n18, 219–20n6. *See also* extensionism; person-centric
*anwēṣaṇam. See* inquiry
Arunima, G., 197–98n62
atheism, 60, 105
attunement: interaction and, 132–34; to pollution, 69–71; suffering and, 209n10
ayurvedic medicine, 215n4

*badalukal. See* alternatives
Bargu, Banu, 152
Baviskar, Amita, 190–91n17, 201n82, 210n16, 214nn41–42
Bellacasa, María Puig de la, 211n21
belonging. *See* community
Benedict, Ruth, 193n25
Björkman, Lisa, 223nn32–33
*bōdham. See* consciousness
*bōdhyam. See* conviction
bodies: as fulcrums for change, 111–13; hierarchy and, 151–52, 189n11, 217n18; pollution and, 69–71, 128, 130–36, 196n43; street protest and, 143–57; vulnerability of, 129, 150–57. *See also* face-to-face interaction
Boggs, Carl, 208n42
bracketing, 73–75, 78
Brahmin castes, 151, 196n41, 214n1, 224n37
*buddhijīvi. See* intellectuals

camps: and environmentalism, 29–30, 39, 53, 56–59, 61, 104; and nature cure, 115, 218n25; as waste of time, 74. *See also* Dialogue Journey; rain camp
cancer, 70, 139, 141, 221n23
Canovan, Margaret, 207n37
capabilities. *See* agency
care, ethic of, 86–87, 211n20
caste: Communist movement and, 60, 203; diet and, 35–36, 202nn4–5, 214n1, 217n18; inter caste marriage and, 59–60, 210n14; localness and, 73–75; markers of, 35, 202n5, 214n1; reform movements, 17, 18, 111, 143, 196n44, 203n12, 217n18; status of activists, 35, 202n5, 210n13
casteism: local identity and, 75, 78, 81, 96; radical activists and, 35, 111, 202n5, 202n4, 214n1; roads and, 43, 143–44, 222n26; subtlety of, 75, 210n14; unapproachability, 196, 222n26; vegetarianism and, 202n4, 214n1. *See also* caste; exclusions
cause: community, basis for, 52–58, 71–73, 208n45; community, opposed to, 3, 5–7, 23, 59, 122–23, 178; community, rooted in, 11–12, 20, 174–75, 192–93n24; community, synergy with, 3, 7–9, 23, 64, 75; interests, opposed to, 21, 62–65, 82–85, 89–90; motive for activism, 2–3, 14, 62–65, 97–98. *See also* living for and living from; purposes
Cavell, Stanley, 195n37
Chandran, Civic, 203–4n13, 204n18
Chatterjee, Partha, 200n75, 207n38
Chaturvedi, Ruchi, 198–99n67, 200n76
Chipko movement, 128–29, 190–91n17, 222n28
Chowdhury, Nusrat S., 45–46, 206–7n34
Choy, Timothy K., 221n19
Chua, Jocelyn Lim, 197–8n62, 198n65
civil society, 21, 22–23, 129, 144, 167, 200n77, 207n38. *See also* democracy
class: and Kerala politics, 17–18, 197n47, 198–99n67; status of activists, 36, 91, 190–91n17, 201n82, 214n42. *See also* Communist movement
Coca-Cola, 9, 15, 102, 112–13, 196n42, 224n44, 225n5
Cody, Francis, 199n71, 206n29, 206–7n34, 219n2
Cohen, Benjamin R., 108, 221n17
collaboration: activist ethics and, 116–17; environmental justice and, 23, 97, 210n15, 214n40; insider/outsider distinctions and, 62–63, 94–96, 97–98, 158, 192n21. *See also* Solidarity Committee
Collins, Patricia Hill, 195n36
common cause: among locals, 64–65, 75, 78, 81; among radicals, 24, 37, 41, 45, 48–49, 123
common people, 5, 25–26, 52, 63, 166, 188n6, 208n44, 215n2
Communist movement, 17–19, 61, 138–39, 196n45, 203n8
Communist Party of India, 17, 38–39, 197n47, 202n2
Communist Party of India (Marxist), 34, 44–45, 198–99n67, 202n2
community: British colonialism and, 21–22, 196n44, 200n78; departure from, 34–42, 105–6, 202–3n7, 203n8; difficulty sustaining, 101, 122–25; formation of, 29–31, 52–61, 71–82; impacted, 64, 71, 82, 201n82, 209n11; Kerala politics and, 19–21, 106; magazines and, 38–41, 57, 61; stance alignment and, 37, 106–8, 121–25. *See also* family; fellowship; localness
community and cause. *See* cause; living for and living from
compromise, 119–122, 123
Congress Party, 17, 74, 148, 153
consciousness: activist inquiry and, 53, 109; consciousness raising, 63, 68, 76, 206n29, 209n8; globe perspective and,

88, 93, 95; as motive, 62, 77, 80, 224n42. *See also* globe perspective; God's-eye view; inquiry
consistency, 122–23, 216n10
consumerism, 188n6, 198n65
conviction, 68–70, 96, 120, 164–65, 168. *See also* consciousness
courts, 10–11, 47–48, 71, 141, 219n2, 221n15
CPM. *See* Communist Party of India (Marxist)
culture, 11–12, 102–3, 193n25, 212n25
Cuomo, Chris J., 211n21

Dalit, 35–36, 202n5, 210n13, 214n1
Das, Veena, 195n37
Daston, Lorraine, 220n14
Davé, Naisargi N., 189n11, 208n44, 219n28
deictics, 136, 220n8
democracy: Kerala political culture and, 17–18, 202n2; in organizing process, 80; radical activists' ideals of, 51, 167; Western ideals of, 21, 167, 199n71, 205n26, 207n38. *See also* people, the; public sphere
departure. *See* community
development, 17–18, 197–98nn61–62, 199n69, 224n45. *See also* Kerala Model; modernity; progress
Devika, J., 12–13, 18, 197–98n62, 224n45
Dhanya: anti-quarry coalition, 32–34, 42–43; criticism of locals, 62–64, 90–91; entry into activism, 34–35; gender inequality, and, 41, 45, 75–77; role in Manamur, 46, 68–69, 78–79
*dhārmikata*, 189
dialogue: awareness raising and, 26, 27, 36, 39, 43, 206n29; inner, 67–68, 98, 109, 177, 225–6n7; objectivity and, 195n36
Dialogue Journey: Ahmed and, 32, 162, 170–71, 172–73, 177–78; history of, 42–43; purposes of, 32–33, 43–44
Diet: caste and, 35, 202nn4–5, 214n1, 217n18; environmentalism and, 85, 100, 108–9, 111, 112–13; vegetarianism, 51, 103–4, 202n4, 214n1, 217n22. *See also* nature cure; tea
directness: ideologies of immediacy, 137–38, 158, 222n25; mediation of, 135, 149, 156–57, 220n14; rage and, 129, 138, 155–56; ritual, 135–37; science and, 142–43; suffering and, 136–137, 139, 154–55, 158–59, 209n10, 221n23. *See also* objects
discussion: little magazines and, 24, 38–39, 39–41; tea and, 109–110, 124. *See also* face-to-face interaction
display: audiences of, 137–38; of dominance, 147–49, 152; mass media and, 44, 129, 149, 222n28; of rage, 146–47, 154–57; social media and, 143–44; streets and, 143–44, 145–46, 152, 157; of suffering, 129, 150–51, 156, 158–59. *See also* performance; street politics
District Collector, 131, 137–38, 144–45, 220n7
Du Bois, John, 106–107, 108, 133–134, 165–166, 216n10
Dumont, Louis, 21–22, 189n11, 199n72
Durkheim, Émile, 193n26, 193n27, 199n72, 226
dwelling, 212n25, 213n37

ecofeminism, 189n9
education, 10–11, 39, 52–53, 193n26, 197n61, 202n6
effects. *See* objects
emotions. *See* rage; sympathy
endings, 160
endurance, 66, 150. *See also* suffering
English medicine, 51, 105–7, 112, 115–17, 124, 215n4, 216n10. *See also* nature cure
environment: body and, 112; meanings of, 87–89, 97, 175
environmental activists. *See* activism; local activists; radical activists
environmental ethics: anti-humanism and, 9–10, 191n18; expansive aspirations, 6, 83, 86–89, 188–89nn7–9, 211n21; non-Western ethics and, 189–190, 188
environmentalism, 188n7, 190–91n17. *See also* environmental justice
environmentalism of the poor, 190–91n17, 191–92n19
environmental justice: alternatives and, 101, 103–4, 111–13; community identity and, 8–9, 88, 90, 97–98, 167, 192n20; concept of, 9–10, 190n15, 190–191n17, 191–192n19; ethics and, 107, 118, 124, 130, 178–79; as form of life, 2, 6, 14–15, 130, 195n39; hybrid structure of, 2, 8, 10, 14, 87, 93, 164, 192n21; mainstream environmentalism and, 9–10, 88–90, 192n20, 206n30; NIMBY and, 64–65, 89–90, 210–211n19; people's protests as, 191n19;

environmental justice *(continued)*
publicity and, 94–96, 128–29; racism and, 9–10, 190n15; related terms, 190–191n17, 191–192n19; science and, 132, 138–39; as uncommon cause, 20–21, 102–5, 128, 146, 148, 199n69; US origins of, 9–10, 190n15, 190n17. *See also* environmentalism; people's protests
epistemic stance. *See* stance
ethical consumption, 112–13
ethics: anthropology of, 12, 193nn26–27, 194n30, 194n32, 212n29, 225n3; definition of, 14, 195n38; forms of life and, 14–15, 194n32; morality and, 189n11, 194n32; moral status, 6, 188nn7–8; impartiality and, 48–49, 83, 86, 89–90, 166–67, 225n2; objects and, 106–7, 130, 133, 136, 147, 157–59, 168; person-centric, 65, 81, 83, 89–94, 158, 167, 212n29, 214n40; pressure and, 12, 58, 106, 114, 120, 124, 168; progress and, 6, 88–90, 167, 188n7, 199n72, 205n6, 212n29; purification of, 23, 120–22, 157; scale and, 87–96, 110–14; self-cultivation, 12, 13–14, 193, 200; subjects and, 129–30, 135, 136, 216; tensions with relationships, 31, 56–61, 105–8, 165–69. *See also* ethic of care; globe perspective; person-centric; stance
ethnography: activism and, 12–13, 24–25, 45–46, 168, 175, 194n33, 201n83; field site selection, 182–83; relationships and, 23–25, 168, 175, 182–3
evidence: legal strategy and, 219n2, 221n15; mass media and, 126–28, 129, 219n2, 222n28; scientific, 132, 134, 137–38, 140–43, 221n19; sensory, 129, 132; social media and, 51–52, 128, 129, 144, 219n2
exclusions: alternatives and, 101–5; consistency and, 114–17; expansive aspirations and, 58, 59–61, 64; localness and, 75, 78–82, 95
exhaustion, 150
experts: experience and, 70, 138; government and, 138, 221n15; solidarity and, 8, 68–69, 221n17. *See also* evidence, scientific
extensionism, 6, 86, 167, 188n8, 199n73. *See also* anthropocentrism; environmental ethics; globe perspective
extremists. *See* radical activists

Facebook. *See* media, social
face-to-face interaction: evidence and, 134–35, 137–38; little magazines and, 24, 38–41; protest tent and, 7–8, 24, 69, 71–72, 144; stance alignment and, 118, 132–137, 217n23. *See also* body; discussion
Faiza: background, 5, 35; choice presented to art students, 4–5, 6–7, 14, 31, 83, 164–65, 187–88, 192–93; concern about Ahmed, 27–28, 163–64; neighbors and, 104–5; relationship with mother, 35, 179
family: departures from, 5, 31, 35–36; ethical differences and, 59–61, 72, 100, 120–122, 179, 219n28; home and, 7, 121–22; opposition to, 118–20; responsibility to, 66, 83–85, 174–75
farming: alternatives and, 9, 37, 51, 56, 99, 122; people's protests and, 44, 133
Faubion, James, 193n27, 194nn30–31
fellowship: camps and, 29–31, 107, 43, 24; freedom and, 57–58; Malayalam *kūṭṭāyma*, 29, 57, 208n43
fish kills, 80–81, 126–27
food. *See* diet
forms of life, 164, 195n37, 199–200n73, 215–16n5
formatting. *See* people, the
Fortun, Kim, 187n3, 201n82, 220–21n14, 224n44
Foucault, Michel, 193n27, 226n9
framing, 139, 204–5nn24–25
freedom: anthropology of ethics and, 12, 167, 193nn26–27, 194n30, 194n32; community and, 60, 105–6, 177; performance and, 96, 130, 158–59; stance and, 106–7, 113–14, 123–24, 167, 177
Freire, Paulo, 206n29
friendship, 4–5, 10–11, 55, 24, 85, 123, 150, 208n45, 210n14
future orientation. *See* localness

Galison, Peter, 220n14
Gandhi, Mohandas K.: experiments, 216n12, 218n26; as moral exemplar, 218n26; nature cure and, 112, 217nn16–18, 218nn25–26; reading and, 207; satyagraha and, 151, 152–153, 223n34
Gatt, C., 208n45, 225n5
gelatin factory: counter-protest strategy, 96, 126, 145; history of, 65–66; pipe leak,

130–131; production process, 209; workers in, xii, 48, 144. *See also* Manamur Action Council
gender: challenging norms, 41, 57–58, 72, 77, 81; inequality in radical activism, 119–120; positionality of researcher, 182, 200; roles in people's protest activism, 45, 75–78, 79, 81, 96. *See also* exclusions; patriarchy; women
Gilligan, C., 211n20
globe perspective, 88, 110–111, 175, 213; anthropology and, 175, 212n29; boundaries of, 219–20n6; encompassment and, 90–91; environmentalism and, 12–13; moral progress and, 88–89, 167, 199n72, 212–13n29; as view-from-somewhere, 93, 213n35
God's-eye view, 88, 212–13n29. *See also* globe perspective
Goffman, Erving, 194n30, 214n39, 215n3
Govindan, M., 38, 39, 42, 43, 57, 205–6n26
Govindrajan, Radhika, 211n21
groups. *See* communities
Guha, Ramachandra, 128–9, 190–91n17, 222n28, 223n34

Habermas, Jürgen, 205–6n26, 207n38
habits, 105, 118, 218n26
Hacking, I., 205n25
Hale, Charles, 201n83
Halliburton, Murphy, 224n42
Hansen, Thomas Blom, 199n70
Haraway, Donna, 195n36, 213n35
Hari: home life, 100; isolation of, 100–101, 108–10, 115; nature cure and, 112, 115–17
Harikrishnan, S., 196n43, 208n43
Harrison, Faye, 12–13, 195n36
Heller, Patrick, 197n54
Herman, Barbara, 224–25n2
highlighting, 74–75, 78
Hofmeyr, Isabel, 207n39
houses, 6, 119, 175–76, 212n25, 226n8
hunger strike, 150–51, 152–53, 158–59

immediacy. *See* directness
impacted community. *See* community
impartiality. *See* ethics
independence movement, 17, 20–21
Ingold, Tim, 87–89, 212n25, 212–13n29, 213n37
inquiry, 24, 52–53, 115, 166–67, 195, 207–8n41
integration. *See* scale
intellectuals: common people, 42, 188n6; little magazines and, 38, 42, 203n13; and Leftism, 203n13; Malayalam term, 217n21
intentions, 82, 214n39
intersubjective coordination. *See* stance
inter-caste marriage, 59–61
interests, 20, 31, 64–65, 83, 89, 98, 139
internal transformation: as aim of activism, 109–10, 111, 163; inscrutability of, 177, 170–72; and performance, 39

Jacob, John C., 39, 42, 43, 204
*Janakīya Samskārika Vēdi. See* People's Cultural Forum
Jeffrey, Robin, 203n8
journalists, 45, 155. *See also* media

Kant, Immanuel, 224–25n2
Keane, Webb, 88, 212–13n29, 220n10
Kerala: casteism and, 196n43, 201n82, 209n5, 210nn13–14, 222n26; economy of, 17–18, 197n61; expansive moral projects and, 17–19, 59, 198n63; geography of, 15–16, 196n44; God's Own Country, 15, 19, 41, 196n41; living for and, 3, 5–6, 17–19; living from and, 20–21; organizations and, 20, 139, 201n82; political culture of, 17–18, 198n56, 198–99n67, 202n2; progress and, 17, 197–98n62, 202n6, 203n13; religious diversity of, 200n76
Kerala Model, 17–19, 197n61, 199n69, 201n82
*Kēraḷa Śāstra Sāhitya Pariṣatt,* 138–139. *See also* people's science
*Kēraḷīyam* magazine: as activist collective, 56–58, 208n43; aesthetics, of, 52; history of, 39, 204–5n24; office, 39–41, 44; people's protests and, 49–52, 206n28; precursors to, 37–39, 204nn18–19, 204n23; protest updates, 46–48; reading practices, 52–53, 207n39; vision for change, 42–43. *See also* little magazines; radical activists
kin. *See* family
King, Martin Luther, Jr., 5–6, 213n38
Kopnina, Helen, 188n7
*kūṭṭāyma. See* camps; fellowship

Laidlaw, James, 193n27, 199–200n73
Lakier, Genevieve, 223n32
Latour, Bruno, 142, 222n25

LDF. *See* Left Democratic Front
leaving. *See* community, departure from
Left Democratic Front, 17–18, 197n47, 202n2, 203n11
legal strategy, 219n2, 221n15
Leopold, Aldo, 24, 188, 200–201n80, 225n5
liberation theology, 221n22
little magazines: alternative to mainstream, 37–38, 103, 203–4n13; discussion and, 24, 38–39, 39–41; exclusiveness of, 42, 203n12; freedom and, 38, 41–42, 61; as institutions, 61; littleness of, 203–4n13; mass media and, 203n12
living for and living from: aspects of human life, 11–15, 22; connected, 166; efforts to harmonize, 2–3, 64–65, 72, 97–98; efforts to oppose, 31, 64, 97, 165–69; not inherently opposed, 22; tensions between, 16
local activists: anti-quarry coalition and, 42–43, 45; combine living for and living from, 2–3, 64–65, 97–98; environmental justice and, 64–65, 90, 192n21; impact and, 71, 209n11; person-centric perspective, 64–65, 90–94; persuading distant audiences, 134–35, 155–56, 158–59, 220–21n14, 223n32, 223n34; science and, 137–143. *See also* localness; Manamur Action Council
localness: attunement and, 69–71, 209n10; bracketing and, 73, 75, 78; care and, 84, 86–87, 211nn20–21; as category of resistance, 103–4, 151; environmental politics and, 89–90; exclusions and, 64, 78–80; not geographic, 80–81, 87, 93, 211–12n23; global and, 90, 209n4; highlighting and, 74–75; impact of pollution and, 67–68, 69, 70, 150; interests and, 20, 31, 64–65, 83, 89, 98, 139; nonlocal affiliations and, 73–74, 148; performance of, 96, 130, 147, 152–53, 156–57, 158–59; person-centric perspective and, 87–88, 95–96; responsibility and, 85–86; solidarity and, 75–76, 81–82, 87, 90–94. *See also* dwelling; person-centric; scale
looping effects, 39, 41, 205n25
Lukose, Ritty, 198n65

magazines. *See* little magazines
Mahmood, Saba, 12, 225–26n7
mainstream. *See* alternatives
Malayalam, xi–xii, xv–xvi, 57, 119, 128, 181
Manamur, 91, 97–98, 210n13. *See also* gelatin factory
Manamur Action Council: caste and, 73, 74–75, 97–98, 210n13; class and, 91, 97–98; Congress Party and, 74, 75, 76, 80–81, 95, 153; gender and, 45, 75–78, 79, 81–82, 96; history of, 65–66; insiders and outsiders, 63, 80, 93–94, 95–96, 97; *Kēraḷīyam* update, 46–49; leadership of, 66, 74, 76, 77, 78, 84, 95; as the people, 46–52, 130, 146–47, 150–60; person-centric values, 90–91; protest tent, 7–8, 12, 24, 25, 69, 71–72, 86–87, 143, 150, 154; street politics and, 143–44, 200n77, 223nn31–32, 233n34; women's committee, 75–77, 78–79, 81–82
Maoism, 38–39, 45, 79–80, 204n19, 206n28, 223n32
maps, 49, 87–88, 91–94, 212n28, 213n35
marginalization. *See* exclusions
marches. *See* street politics
Martellozzo, Nicola, 220n6
Martinez-Alier, Joan, 190–91n17
masses, the. *See* people, the
Mattingly, Cheryl, 193n27, 216n12
Mazzarella, William, 135
McGurty, Eileen, 89–90
McKee, Emily, 213n37
media: events and, 128, 219; newspapers, 44, 52, 103, 129–30, 159, 219n2; perspective on protest, 221–22n24; publicity and, 129, 222, 150–51, 153; social media, 144–45; street politics and, 147–49, 200n77; TV and, 126, 128, 142, 143, 143–44, 156–57, 184, 219n1
Menon, Dilip, 196n45, 203n8
meta-ethics. *See* activist ethics
Mitchell, Lisa, 200n7, 203n33
modernity, 18, 167, 197–98n62, 206–7n26, 218n26. *See also* Communist movement; progress
modern medicine. *See* English medicine
moral circles, 6, 89, 212n29. *See also* environmental ethics; extensionism; scale
morality. *See* ethics
moral objects. *See* objects
moral status. *See* ethics, moral status
moral trailblazers, 5–6, 218n26
moral worlds, 133
Morton, Gregory, 202–3n7
*mūlyam,* 189n11
mundane. *See* ordinary

Næss, Arne, 88–89, 189n9
Namboodiripad, E. M. S., 197n47
Namboodiri caste, 203n8
NAPM. *See* National Alliance of People's Movements
Narmada Bachao Andolan, 195–96n40, 199n69, 204–5n24
*nāṭ*, 63–65, 68–69, 187n2, 190n14, 208n1. *See also* localness; shifters
National Alliance of People's Movements, 203n9, 204–5n24
nature cure: alterity and, 113–17; camps and, 115, 218; freedom and, 114, 123; Gandhi and, 112, 217nn16–17; historical roots of, 218n26; moral pressure and, 113–14, 123; nature life and, 112–13, 115–17, 218n25; naturopathy and, 103–4, 217n16; ordinary and, 215–16n5; scale and, 110–11. *See also* alternatives
nature life. *See* nature cure
naturopathy. *See* nature cure
Nayar caste, 65–66, 197–98, 203n8, 209n5, 210n13
NGOs, 6, 104–5, 139–40, 163, 178, 181–82
NIMBY. *See* Not-In-My-Backyard activism
nonhumans, 6–7, 189, 211
nonlocal. *See* localness
normal, 104, 163, 215n2, n3
norms: breaking of, 7–8, 210n15, 216n6; freedom and, 7–8, 12–13, 124, 166–67, 193n27, 226n7; making of, 58; morality and, 189n6, 194n32. *See also* ethics
Not-In-My-Backyard activism, 62–65, 83, 88–90, 210–11n19. *See also* localness; person-centric

objectification. *See* objects
objectivity. *See* objects
objects: evidence and, 132; humans and, 130, 219n4, 219–20n6; intersubjective coordination and, 133, 135, 142, 220n9; objectification and, 130; objectivity and, 142–43, 195n36, 220–21n14, 222n25; object-oriented ethics, 136, 167–68; the people as, 206–7nn34–35; performance and, 214n39; spontaneity and, 156, 203n30, n32; stance and, 107, 165–66, 216n10; sympathy and, 159, 206–7n34; unquietness of, 129–30, 137, 142, 219n5; vulnerability of, 147, 150, 153. *See also* directness; ethics; stance

One World University, 108–109, 114–15, 123
ordinary: ethics, 215–16n5; people, 25–26, 52, 63, 166, 188n6, 208n44, 215n2; things, 5, 102, 215–16n5
organizations, 10, 19–20–21, 22, 48–49, 208n43
Osella, Caroline, 197–98n62, 198n65
Osella, Filippo, 197–98n62, 198n65
Ottinger, Gwen, 221n17
outsiders, 63, 80, 93–94, 95–96, 97. *See also* exclusions; localness; solidarity

Pañcāyatt, 72, 75–76
Pandian, Anand, 200n74
parallel media. *See* alternatives
parenting, 106
partiality, 48–49, 224–25n2. *See also* impartiality
party politics, 34–35, 37, 59, 60–61, 203
*Pāṭhabhēdam*, 38–39, 204n18
patriarchy, 77–78, 118–20, 210n15. *See also* gender
people, the: camps and, 36, 57, 167, 218n24; formatting of, 46–48; Maoism and, 39, 44–45, 206n28; masses and, 138–39, 199n70, 207n35; as missing, 45–46, 193n27; as object, 206–207n34; performance of, 96, 130, 147, 152–53, 156–57, 158–59; persuasive power of, 132, 135, 137–38, 158–59, 167–68; public and, 96, 130, 143, 153; reactive, 82, 156; street politics and, 143–44, 200n77, 222n27, 223nn31–33; thickening, 49–52; thinning, 48–49, 207n38; as universal, 48–49, 51. *See also* people's protests; people's science
People's Cultural Forum, 204n19
People's Planning Campaign, 18
people's protests: alternative media and, 37–38, 103, 203–4n13; political parties and, 20–22, 203n9; public approval of, 20–21; solidarity with, 23, 42, 43–44, 46–47, 190n13; street politics of, 200n77, 222n27, 223nn31–33; as uncommon cause, 20–21, 102–105, 128, 146, 148, 199n69. *See also* Manamur Action Council; people, the; quarry coalition
people's science: bias and, 140–41; collaboration and, 140–41, 137–38, 221–22n24; emotion and, 140, 155; facts and, 140–42

performance: experience and, 138, 139; power and, 145–47; silence and, 138. *See also* people, the
person-centric, 63–65, 81, 83, 90–96, 212n29. *See also* ethics; local activists; localness
Plachimada, 15, 112, 196n42, 224n44
police, 62–63, 71, 84, 149, 151, 153, 154, 201n84
Polletta, Francesca, 156, 214n39
pollution: attunement to, 69–71; evidence of, 131–32, 139, 14–41, 142–42; experience of, 64–65, 160, 134–35, 158–59, 209n10, 221n23; gelatin and, 7–8, 9, 85, 128, 130–31, 160; health effects of, 68, 70, 150; mediation of, 128, 130–32, 135–36, 142–43; as motive, 80–81, 83; sensory experience of, 129, 132–133, 135–36; as the smell, 65–66, 130–32; social effects of, 67–69; speaks for itself, 137, 142, 222n25; sympathy and, 137–38; tours of, 158–59, 129, 133–34, 135, 136–37, 220n12. *See also* directness; objects
power politics, 150
*prakṛti cikilsa,* 100, 112
Prasse-Freeman, Elliott, 195–95n33
prefigurative politics, 208n42, 216n13
*Prēraṇa,* 38–39, 204n19
primitive, 21–22, 89–90, 199n70, 212–13n29
progress: ethics and, 88–90, 167; globe perspective and, 87–88; Kerala and, 18–19, 197n70, 197–98n62, 199n69
protest: bodies and, 44, 129, 143–144, 146–47, 223n32, 224n37; CPM and, 145, 47, 147–49, 197n47; opposition to, 19–20, 198n65; as performance, 147, 152–53; prevalence in Kerala, 17, 18–21. *See also* people's protests; street politics
protest tent: description of, 7–8, 24, 71–72, 144–145; local identity and, 75–76, 85–86; the people and, 44, 143–145, 150, 160; stance and, 12, 24, 69
public sphere: as split, 200n77; street politics and, 196n43, 222n27, 223n32; Western ideals of, 21, 23, 167, 199n71, 205n26, 222n27
Pulido, Laura, 190n15
purposes: activism and, 11–13, 24–26, 168; as choices, 3–5, 192–93n24; communities and, 42, 61; humans and, 11–12, 192n22; interests and, 20, 31, 64–65, 83, 89, 98, 139; larger, 31; performance of, 214n39; tension with relationships, 3, 64, 194n32. *See also* cause; ethics; living for and living from

quarries: Ahmed and, 1–2, 169, 171–76; environmental effects of, 43, 169, 170; mafias and, 174, 175–76, 225nn3–4, 225n6; protests of, 32–33, 46, 49–51, 218n27

radical activists: alterity of, 36–37, 58, 61, 102–4, 107; caste and, 6, 35–36, 69; class and, 36; different thinking, 5–6, 24, 38, 51, 100; discussion, 24, 38–39, 39–41, 109–10, 124; entry into activism, 34–37, 41–42, 54–56; formation of, 52–53; freedom and, 12, 123, 41–42, 106; genealogies of, 37, 39; gender and, 51, 57–58, 119–20; itinerant, 43–44; magazines and, 3–4, 51; nature cure and, 110–11, 112, 113–14, 115–17, 218nn25–26; new kind of community, 29–31, 56–61; not the people, 8–9, 46, 60–61, 91, 102; organizations and, 39, 61, 202–3n24; recruitment, 52, 55–56; solidarity with the people, 6–7, 43–44, 190n13; values over relationships, 31, 56, 59, 61, 101, 163–64. See also *Kēraḷīyam* magazine; Solidarity Committee
rage: irrationality and, 155–56; objectivity and, 220–21n14; performance of, 156–57; pollution and, 155–57; as strength, 154–55; violence and, 127, 154–55
rain camp: belonging and, 59; difference and, 37; the people and, 61; song and, 29, 30–31; unity and, 36, 100–101
raising awareness. *See* consciousness
reading, 17, 31, 53, 207n39
recruitment, 52, 54–56, 66–69, 152
religion: ethics and, 2, 193n26; marriage and, 9, 35, 59–60, 104–5; universality and, 59
responsibility, 85–87, 209n10
roads, 143–44, 222n26. *See also* street politics
Robbins, Joel, 214n40
Roy, Arundhati, 196n41, 199n69
Roy, M.N., 38

*sādhāraṇa. See* ordinary
*sādhāraṇakkār. See* common people

Sahlins, Marshall, 212–13n29
*samaram*. *See* protests
Sartre, Jean-Paul, 192–93n24
*satyagraha*, 151, 152–53, 223n34
scale: breadth and, 88–89; collaboration and, 89–90, 94–96; configuration and, 81, 112–13, 211–12n23; environmental politics and, 87–88, 211–12n23; ethics and, 88–89, 91; integration, 51, 91–92, 111; interscaling, 213n31; nature cure and, 110–11; performance and, 94–96, 214n39; positionality and, 90, 96. *See also* globe perspective; localness; person-centric
schools, 52–53, 104, 198n66, 202n6, 207n40
Schutz, Alfred, 133, 220n9
SEEK. *See* Society for Environmental Education in Kerala
self-cultivation. *See* ethics
self-interest. *See* interests
self-reflection, 176–77. *See also* consciousness
sensory experience. *See* pollution
Shah, Alpa, 194n33, 226n5
Shapiro, Nicholas, 209n9
shared moral reality, 132–34. *See also* stance
Shiva, Vandana, 189–90n12
Sidnell, Jack, 217n22
Singer, Peter, 191n18, 199n73, 224–25n2
Singh, Prerna, 198n63
silencing, 75, 95–96, 100, 175. *See also* exclusions
smell, 78–79, 208–9n2, 209n6. *See also* pollution
Smith, Adam, 194n30, 217–18n23, 224–25n2, 226n9
Smith, Kimberly K., 190n15
SNDP. *See* Sree Narayana Dharma Paripalana
social media. *See* media
social work, 1, 12, 26, 27, 161, 178, 188n7, 189n9, 212n26
Society for Environmental Education in Kerala, 39, 204n23
solidarity, 190n13, 221–22n24
solidarity activists. *See* radical activists
Solidarity Committee: Dhanya leadership of, 23, 62–63, 33–34; failure of, 46–47; roles of, 43–44, 190n13. *See also* radical activists
songs, 29–30, 6, 32, 33, 39, 56–57, 105
speciesism, 188–89n8, 191n18. *See also* anthropocentrism; extensionism
sphere. *See* dwelling; person-centric
Spivak, Gayatri, 206–7n34, 219n5
spontaneity. *See* objects
Sree Narayana Dharma Paripalana, 217
Sree Narayana Guru, 111, 217
stance: consistency and, 216n10; epistemic, 133–34; ethical, 165–66, 219n28; objects and, 133–34, 135–37, 156, 219n4; subjects and, 129–30, 135, 136, 216n13; preference for alignment of, 107, 113–14, 117, 123, 216n10; relationships and, 107–108, 122–23, 182–83; triangle, 106–7, 108, 166
street politics: bandhs, 223nn32–33; caste reform movements and, 17, 143; crowds and, 222n27, 223n31; hartals, 147–148, 223nn30–34; marches, 1–2, 19, 71–72, 222n26; mass media and, 129–30, 149; people's protests and, 144–45, 200n77, 223n31; political parties and, 145–46, 147–49, 152, 223; publicity and, 143–144, 149, 222n27; roadblocks, 69, 152, 223n32; social media and, 129–30, 219; the state and, 148, 223nn30–33; violence and, 149, 154, 157, 158–59
subaltern, 191–92n19, 200n77
Subaltern Studies, 206n34, 219n5
subject formation. *See also* ethics; stance
subjectivity, 130, 136, 157–58, 159–60, 219–20n6, 220n9
suffering: attunement and, 69–71, 209nn8–10; display of, 129, 150–51, 156, 209n10. *See also* endurance; pollution
*Sūcimukhi*, 23
Sunitha: electoral candidate, 72–73; entry into activism, 2–3, 8–9, 10–11, 66–68; future plans, 85–86; as mother, 8, 12, 66–68, 84; solidarity activists and, 68–69
Sunny: field site selection and, 183; *Kēraḷīyam* magazine and, 51–56, 59–60, 101–2; nature cure and, 116–17; rain camp at, 29, 36, 39, 56–57, 100
Sylvia, Tesh, 138
sympathy, 137–38, 159, 206–7n34

Taylor, Charles, 192–93n24
tea, 100, 102, 104, 109–110, 113–14, 118, 216n11
tent. *See* protest tent
thickening. *See* people, the
thinning. *See* people, the

Thiranagama, Sharika, 210n14
Thomas, Sonja, 196n41, 198n66
Throop, C. Jason, 209n10
transcription, 182, 184–85
translation, 15
tribal groups. See *adivasis*
Tronto, Joan, 86, 194n30, 211nn20–21, 224–25n2
Tsing, Anna, 87, 211n22
TV. *See* media

unapproachability, 222n26. *See also* casteism
uncommon: environmental justice as, 20–21, 102–5, 128, 146, 148, 199n69; radical activists as, 6–7, 36–37, 61, 101–4, 107. *See also* alternatives; people's protests

values: and ethics/morality distinction, 189n6; interests and, 31, 83–84; over relationships, 31, 56, 59, 61, 101, 163–64. *See also* cause; ethics
Vanchi Lodge, 38–41, 42, 45, 203–4n13, 206n28
Vākk, 204n18
VapoRub, 117–18, 163, 215–16n5
vegetarianism. *See* diet
Venkat, Bharat Jayram, 218n26
Venkatesan, Soumhya, 194n30
video recording, 44, 127–28, 132, 181, 184
Vijayan: health, 71; leadership of Action Council, 78–80, 84–85; party politics and, 74–76; protest update and, 46–48; Solidarity Committee and, 95–96
vikāram. *See* rage
Vimōcana Samaram, 198n66
violence, 62–63, 127, 154–55, 157, 158–59
Voluntary Human Extinction Movement, 191n18

Weber, Max, 11, 192n22
Whyte, Kyle Powys, 211n21
Williams, Bernard, 191n18, 195n38, 210n17, 224–25n2
Wittgenstein, Ludwig, 14, 195n37
women: Kerala model and, 18, 210n15; motherhood and, 8, 12, 66–68, 84; roles in people's protests, 75–77, 81–82, 129
women's committee. *See* Manamur Action Council
Wright, Fiona, 190n13

Zigon, Jarrett, 193n27
Zuckerman, Charles, 13

www.ingramcontent.com/pod-product-compliance
Lightning Source LLC
Chambersburg PA
CBHW020844270326
41928CB00006B/545